The Dark Sides of the Internet

ROLAND HEICKERÖ

THE DARK SIDES OF THE INTERNET

On Cyber Threats and Information Warfare

PETER LANG

Frankfurt am Main · Berlin · Bern · Bruxelles · New York · Oxford · Wien

Bibliographic Information published by the Deutsche Nationalbibliothek
The Deutsche Nationalbibliothek lists this publication in the Deutsche Nationalbibliografie; detailed bibliographic data is available in the internet at http://dnb.d-nb.de.

Cover and Photo Design:
© Olaf Glöckler, Atelier Platen, Friedberg

Translated from the Swedish original
Internets mörka sidor. Om cyberhot och informationskrigföring,
published by Atlantis Bokförlag, 2012.

Translated by Martin Peterson.

ISBN 978-3-631-62478-4

© for the English edition: Peter Lang GmbH
Internationaler Verlag der Wissenschaften
Frankfurt am Main 2013
© for the Swedish edition: Bokförlaget Atlantis AB, Stockholm
All rights reserved.

All parts of this publication are protected by copyright. Any utilisation outside the strict limits of the copyright law, without the permission of the publisher, is forbidden and liable to prosecution. This applies in particular to reproductions, translations, microfilming, and storage and processing in electronic retrieval systems.

www.peterlang.de

Table of Contents

Preface 7

Introduction 9

Players, means and methods 19

The Superpowers 31

Cyber terrorism. Electronic Jihad 59

Industrial espionage and theft of information 77

Cyber crime 95

Aggression between hackers 117

Towards a new modus operandi? The Estonian and Georgian cyber war experiences 131

The need for a treaty in cyberspace 141

Future cyber threats 147

Conclusions 155

References 157

Preface

Years ago, when I worked in the telecom sector for the wireless operator Telia Mobile and later on for Ericsson, I became interested in the mega change we are into to today – the information revolution. At that time mobile phones and the Internet were relatively new phenomena. As more and more people are getting connected to the global network and use more advanced services and applications, questions arise regarding how to secure and protect information, and information and communications systems.

From having developed mobile systems, I now try to understand vulnerabilities and security holes in networks and functions for illigetimate cyber activities and their eventual consequenses at security policy level as well as at socio-technical systems levels. This book tries to explain the growth of varied and serious phenomena on the net, the dark sides of the Internet, including the battle for information superiority, cyber terrorism, criminality and espionage as well political-ideological hacking, "hacktivism".

For fruitful discussions on different aspects of information security and cyber threats, I would like to thank my friend and co-worker Dan Larsson, senior expert at the Swedish Defence Radio Establishment. I also want to express my gratitude to "Nico" and his colleagues at the Swedish Armed Forces for their profound knowledge and stringency within the field. Lars Nicander, head of the Center for Asymmetrical Threats and Terrorism Studies (CATS) at the Swedish National Defence College, deserves great praise for his expertise and his neverending will to convey insights and knowledge on this important subject.

A also want to give my thanks to my former collegues at the Swedish Defence Research Agency, especially John Rydkvist, Jerker Hellström and Alexander Atarodi.

This book is dedicated to Anna, Vilgot, Hanna and Moa.

Stockholm, September 2012

Introduction

Terms like cyber threats, information warfare and digital attacks are charged to many people, surrounded by an aura of mysticism and secrecy. This is quite natural. The phenomena are relatively new. They are a result of the information revolution, which in its turn can be experienced as complex, difficult to interpret and multidimensional. It is also a field surrounded by a high level of secrecy, since it involves different kinds of illicit influences on information and information systems – and protection against such activities.

The first time the word "cyber space" was mentioned in more broad terms was in the 1984 novel *Neuromancer* by science fiction writer William Gibson. It has become very popular and is used in a number of different contexts to describe the virtual world of online computers and networks and what is in between in the form of data flows and interacting users.

Pioneering technological breakthroughs such as the Internet, mobile phones and computers have resulted in radical changes in people's possibilities to communicate and spread information as well as handle electronic transactions of different kinds. These innovations are just as important and decisive to humanity as Gutenberg's printer in his days. They fundamentally affect societal structures and give rise to new logics and business models. The Internet makes it possible for users to gather, store, process and send large amounts of data. It is a force that makes globalisation possible. The Internet breaks up information monopolies and opens up for competition about the scope in cyber space.

Access to distribution channels such as TV, radio and newspapers are no longer as vital as they have been to spread messages and influence public opinion. Via blogs, web sites and Internet forums practically anybody can make his or her voice heard. This changes power conditions. Increased opportunities to communicate and spread information also means that new technology and new methods, which previously were difficult to provide or too expensive and complex, can be used by more and more people.

The use of communications technology penetrates all levels of society. In high-technological societies it can be difficult to defend oneself against the wave of data that flows from apps, cell phones and computers, and which often demands that the user is available immediately. A lot of the information that is spread can be perceived as relatively uninteresting, other parts as important and valuable. It can lead to increased understanding and a better grasp of various phenomena. Attitudes and outlooks change. In the long run this can lead to increased democratisation and new knowledge.

Non-democratic regimes hardly see the development as positive, but rather as a threat. The powers that be try different methods, such as censorship and supervision of users, to prevent information from being spread. They obstruct the conditions to create meeting places – communities – on the web where knowledge and experiences can be conveyed. The capacity to process and convert data into information and simply and cheaply send it to third parties sets powerful forces in motion. In extreme cases it can manifest itself in popular protests against abuse of power such as in North Africa and the Middle East in the spring of 2011 and in Iran in 2009.

The global number of users with access to the Internet is more than two billion. Every month the web grows by tens of millions of new users. Since its commercial breakthrough in the mid-1990s the Internet has developed into an enormous meeting place and linking force. Cyber space can be defined as multi-polar in the sense that it consists of a countless number of players from all the corners of the world. Using different kinds of information and communications services has become part of everyday life in all modern societies and most people can now hardly imagine life without them. Social media such as Twitter and Facebook are used to connect people. Phenomena such as Wikileaks, Wikipedia and YouTube would not exist without the global web.

While information technology has a number of positive results for most people, harmful conducts and activities have also evolved parallel to these. The conditions for antagonistic conduct have changed. New threats are created as cyber space expands. This kind of aggression cuts through the whole societal spectrum and affects both civilian and military structures, governments, organisations and companies as well as the individual citizen. The reason for this is partly society's major dependence on information technology, partly that the systems are so intimately intertwined. Cyber attacks against the electrical grid, for example, can have an effect on telephone and computer networks. This in its turn risks having dispersion effects in other sectors and businesses such as financial systems, aviation and so on. The result is serious disruptions of infrastructure that is vital to society.

Information warfare

Information warfare can be used at strategic political-economic level in diplomatic talks between parties during negotiations or as an economic means of bringing pressure to bear on another nation. It can also be used at tactical and operational level close to or in connection with a military conflict.

However, information warfare is in itself not in any way limited to states' actions and relations between nations, but covers the whole range of human activities and conducts. In the original sense it is referred to the individual and takes place in interaction between people and organisations. This has been going on for as long as humans have existed. The purpose is to make opponents act in a way that is advantageous to oneself. Human perception can be manipulated in subtle ways at which the subject's opportunity to correctly interpret the situation is rendered more difficult. This can either be done directly against the opponent, eye to eye, or indirectly via different kinds of technical systems and means.

Access to information and the ability to use it to one's own advantage are intimately related to power and control. On the analogy of this, the lack of relevant information or access to inaccurate information results in an incapacity to make correct decisions, which may be decisive and, depending on the context, lead to serious consequences. This applies generally, irrespective of whether it involves a person who is going to perform a single transaction, a company which is going to make an investment decision or a great power that bases its strength on information superiority.

Information warfare is a comprehensive collective term which includes different parts of the electromagnetic spectrum such as radio, radar, laser and electromagnetic pulse weapons. Other activities that belong to this category are psychological operations, military deception and computer and network operations. The latter are usually designated cyber warfare on the Internet. Included in information warfare are also different aspects of protection of information and systems.

Computer and network operations are asymmetric in character. From a relatively minor stake, the effect can be very big. A digital attack against a system requires fairly small resources, while it is costly to defend oneself. That way, the barriers are lowered for this kind of attack.

Using modern information technology, the distance to the target is not necessarily a problem. Cyber threats can arise anywhere in the world and knock out not only computers but cell phones and other units that communicate over the Internet. In some cases it is difficult to know who the actual instigator behind an operation is. The same can also be true of the target and the purpose of it. The web makes anonymity possible. Intentions can be hidden in the cyber fog.

At the same time, anonymity means that there is a risk that the wrong culprit is singled out and that defence measures against an attack are disproportionate – that the defending party resorts to more advanced cyber weapons against the party that is believed to have initiated the attack. That way, there is a risk that a conflict escalates very quickly.

Examples of targets are individual states, but also companies and organisations, groups and individuals. Harmful cyber activities can involve spreading propaganda or misleading information, industrial espionage or aim at tapping, interrupting or destroying information, computers and networks. The opponent can also have his access to vital and sensitive information blocked and thus be prevented from acting.

The Internet is in itself a very useful channel for distortion through rumours and slander conveyed *en masse* via social media and networks. Anyone exposed to this will find it very difficult to defend him- or herself. Just as the Internet is a channel for dissemination of information, it can be used for repressive purposes by less scrupulous regimes and organisations in order to survey the activities and conduct of their critics on the web. By e.g. planting malicious code, such as viruses, worms and trojans, infringements and hacking as well as denial-of-service attacks against important systems, an opponent's desire and capacity to defend himself can be reduced.

In cyber space there are no clear boundaries between military and civilian infrastructure; they are integrated and interdependent, since they often share the same communications networks. As long as nations depend on computer networks for their basic economic and military power, and as long as these can be accessed from the outside and the inside by people with bad intentions, there are thus risks.

An advanced cyber attack directed against an opponent's vital information infrastructure – especially command and control functions in electricity, telecommunications and computer systems, financial institutions, the energy sector, air traffic etc. – can have serious social and security political effects in a short time. Important and vital functions may be discontinued permanently or over a limited period of time. This means great challenges to the responsible parties.

The first officially published cyber attack on an individual state was carried out against Estonia in the spring of 2007. In connection with the transfer of a Soviet military statue from central Tallinn to a cemetery nearby in the city, a number of cyber attacks were initiated against computer systems in the Estonian parliament and several ministries, two of the biggest banks and six news organisations and various other national web sites. In a few days servers and networks were overloaded, at which functionality went down. Who or what groups and organisations were behind this attack has not been made clear. But suspicions have been pointed at Russian nationalist hacker groups – so-called hacktivists. The attacks ceased as quickly as they began.

In a 2003 intelligence report on cyber defence by the U.S. Navy the following estimate was done: A group of some thirty hackers, strategically located and with a budget of less than 10 million U.S. dollars, could shut down large parts of the critical infrastructure in the United States in a well co-ordinated attack.[1]

A computer found by U.S. forces in one of al-Qaeda's buildings in Kabul contained a model of a virtual dam.[2] Apart from the model there was a programme that simulated what effects there would be on people and equipment if the dam broke. The direction and propagation of the water masses could be calculated. The example shows that al-Qaeda's activists are well versed in the use of computers during tactical planning of operations.

---

1   Atwan, A.B. (2006) "The Secret History of al Qaeda". *University of California Press*. Berkeley, USA.
2   Borger, J. (2002) "US fears al Qaida hackers will hit vital computer networks". *The Guardian*, 28 June 2002.

Cyber aggression

Cyber warfare is something that is pursued by many kinds of players. Behind an operation there may be a resourceful national intelligence and security service as well as an organisation without national ties and which acts independently and idealistically. An operation can also be carried out by a small group of dedicated hackers with good IT skills, by criminal organisations, by insiders and terrorists. Unholy alliances can arise between different kinds of players. For example, a state can use cyber crime groups as tools to attack another state.

For the great powers, cyber space is an area of battle and protection. It is a question of defending the country's sovereignty in the digital sphere. Cyber weapons can be used as an individual resource against an enemy or in combination with conventional weapons systems such as aircraft, tanks, missiles and weapons of mass destruction. It changes the balance of power between opponents, since a party that is militarily weaker on paper can gain strategic, operational and tactical advantages in cyber space against a seemingly stronger rival. This makes cyber warfare attractive.

Planting malicious code such as the *Stuxnet* worm, which was discovered in June 2010, is an example of this new phase of warfare.[3] The malicious code was tailor-made to attack master computers at the Iranian uranium enrichment facility in Natanz. Through Stuxnet, centrifuges that were used for enrichment could be inaccurately adjusted which resulted in a number of them being subjected to abnormal wear and destroyed. In different contexts Israel and the United States have been singled out as the instigators of the operation. In 2012 an even more advanced malware was discovered known as *Flame* or *Skywiper* addressed to attack computers running Microsoft Windows. It targeted computers in the Middle East, especially in Iran, and some sources claim that Israeli computer experts were behind it. In the autumn of 2012 another malware, *Gauss*, was discovered – it is similar to Flame and designed to hit banks in the Middle East. The virus is said to come from the same "factory" as Stuxnet. Other examples of events that can be described as cyber warfare include Chinese hacker groups that have broken in to the Pentagon's computer systems in order to steal sensitive information.

---

3   Beaumont, C. (2010) "Stuxnet virus: worm could be aimed at high-profile Iranian targets". *The Telegraph*, 23 September, 2010.

Today all technologically advanced states are developing concepts for network and computer operations. Doctrines are rewritten and adjusted to altered demands and new threats. The great powers are setting up organisations in order to defend their own information resources and communications systems and be able to spy on and strike against hostile ones. In October 2010 the *U.S. Cyber Command*, specialised in digital warfare, became operational. The organisation is led by the chief of the strategic signals intelligence organisation, the NSA.[4] As of this, cyber warfare has been ranked in the same category as the other strategic functions for the army, navy, marine corps and air force.

In a similar way Russia and China are building up their capacity, resources and units for digital warfare under the command of their military organisations and intelligence and security ministries. NATO, in its strategic doctrine for 2010, specifically pointed out cyber warfare as one of the most serious threats to the alliance and the countries that are part of it.

Western European countries are in no way exempt from antagonistic activities that can be suspected of having links to foreign powers. Foreign hacker groups have made attempts to penetrate objects of vital importance to society. High-technological activities are very much exposed to cyber espionage. This is especially serious for a country that builds its economic base on innovations. Cyber crime directed against companies and individuals is a growing problem, with huge sums in yearly turnover.

Terrorism over the Internet, where web pages and other means of communication are used for spreading messages and radicalising people, has been performed by European citizens. One conspicuous example is the case of the Swedish suicide bomber Taimour Abdulwahab in Stockholm a couple of days before Christmas 2010. In the wake of the attacks by Islamist Mohamed Merah in Toulouse in March 2012, French authorities have talked of banning its nationals from visiting web pages on the Internet that harbour radicalising messages.

Structure and contents

The cyber attack against Estonia in the spring of 2007, when a number of servers and networks were overloaded, shows the possible effects and conse-

---

4   NSA – The National Security Agency.

quences of a computer and network operation. The case has become something of an alarm bell when it comes to the risks of attacks against vital communications infrastructure. The great powers building up their cyber warfare capacity, the Stuxnet worm and hacker operations in order to steal information reflect the same reality. Added to this there are various kinds of illegitimate and criminal activities.

Every modern state thus has to create strategies and courses of action in order to protect information, networks and computers that are vital to society from malicious cyber activities. Creating secure systems and minimising risks of information being leaked or tampered with should be a prioritised task. It is also important to understand what threats arise from the information revolution.

The purpose of this book is to give a broad background to the development of the dark side of the Internet and its consequences. It is not about scaremongering, but about creating understanding and knowledge and thus preparedness in order to handle detrimental activities. It describes the change in progress and what it may mean to society, companies and individuals as well as to military and police activities. Important issues such as integrity and censorship and the like are not specifically discussed in the text, only in general terms. Nor can every aspect of information security be described.

The first chapter of the book initially describes different kinds of players and antagonists with malevolent behaviour on the Internet. This is followed by a discussion on the methods used for detrimental activities via computer and network operations. The text gives a general background to the coming chapters in the book, where the player categories and different phenomena are discussed in more detail.

The second chapter of the book deals with the American, Russian and Chinese view on information warfare, which has been described in different doctrines and by leading military theorists and analysts. There is a historical background from the 1970s and onwards as to how different concepts and theories have developed from command warfare and network centric logic to information operations and cyber warfare.

Cyber terrorism is a phenomenon that is gaining more and more attention. One reason is the concern that modern information and communications technology may be used in order to harm open societies. This concern also involves actual IT systems and the information generated being targets of advanced attacks. That way functions that are important to society could be affected. The term cyber terrorism is complex. The third chapter of the book describes the difference between traditional and cyber terrorism. The main focus is on how the al-Qaeda terrorist network acts in cyber space, and on al-Qaeda's change in concentration and activities into a clever player in an electronic Jihad.

One of the most serious threats to a modern country's trade, industry and long-term economic development is industrial espionage and insiders. The activities are directed against high-technological industries and companies with advanced basic research. The defence and telecoms sectors are of particular interest, just as biotechnics, medical and material technology. Behind this kind of espionage there may be individual states and security services as well as competing companies. One trend is that criminal players are getting involved both as thieves and fences of information. Computerisation and the development of the Internet drastically increase the possibility of procuring sensitive information through illegal means. This can be done in different ways. In the fourth chapter there is a background. A number of examples are provided of how industrial espionage has been revealed and of the methods used during collection of information over the Internet – such as signals intelligence, monitoring of traffic, penetration and overtaking of computers with the aid of trojans. The chapter ends with a discussion of the Chinese cyber espionage network GhostNet which was discovered in 2009.

Criminality over the Internet is a growing problem. Globally, huge sums are being turned over every year. The fifth chapter of the book describes the development of small-scale digital criminality in the early 1970s to the present advanced, large-scale and cross-border operations. Examples are provided of the actions of criminal groups and hackers on the dark side of the Internet in order to steal information and commit fraud. The chapter also describes the black market for stolen information, which is traded and sold on secret servers. A lot of the crimes can be related to areas with a lot of corruption in combination with weak legislation in defence of private property. Quite a lot of cyber criminality derives from Eastern Europe, South East Asia and parts of Africa, particularly Russia, China and Nigeria. American cyber crime is also discussed; some of the world's most notorious hackers are from North America.

Hacktivism – or politically, religiously, ethnically and ideologically motivated hacking of opponents' networks and servers – is a growing problem. The sixth chapter provides several examples of hacktivism that have led to confrontations between groups of hackers and where the origin lies in political, religious, national and ideological differences of opinion. The text discusses the confrontation between Palestinian and Israeli groups, conflicts between Islamist and Western hackers, nationalist chauvinist manifestations between Pakistan and India as well as Chinese and Russian activism on the Internet. Finally, there are some conclusions about the risks of cyber escalation and possible security policy implications.

When it comes to hacktivism directed against individual states, companies and organisations, two events have above all been paid a great deal of attention in the last few years. The one is the digital attacks against Estonia in 2007; the other is the operation against Georgia in 2008. Both examples give an indication about the way future cyber operations may take place. The events show a paradigmatic shift; net activists either act in their own interest or as proxies for one party in a military conflict.

The seventh chapter initially describes the Estonian cyber conflict and then the combined military and digital operation against Georgia. The events are discussed and analysed, and a number of conclusions are drawn.

Cyber attacks are one of the greatest modern threats to international peace and security. Making sure cyber space is secure is of the utmost importance. Development of different kinds of antagonism on the Internet and leading nations' build-up of capacity for cyber warfare brings the need for regulation of detrimental and illegitimate activities to the fore. There are a number of areas of sensitive character that have to be tied to international conventions and attitudes. Naturally, the need for cyber regulation has to be handled with consideration paid to aspects such as democracy, openness and freedom on the Internet in relation to things like risks of unacceptable supervision and control, censorship and monitoring. The United States, Russia and China have differing points of view on the issue and present different arguments when it comes to the need for regulation. Chapter eight describes these countries' attitudes and perspectives.

In view of the previous chapters in the book, the ninth and final chapter discusses the change in cyber threats over time. There is a brief summary about the possible future development of different phenomena in cyber space.

Players, means and methods

Just as in the physical world, most of us behave well and legally in cyberspace, but some people – quite a few looking at the great number of users connected to the Internet – act as criminals from a judicial point of view. They conduct felonious and/or unethical acts in one sense or another.

According to the Internet security firm Symantec, crimes on the net occur as often as every third second. This involves everything from information theft, fraud and blackmail to hacker attacks. Cyberspace also makes up an area of conflict between individuals, groups and nations, with true digital battles. There is a wide range of malicious players; from states and security services to cybercrime organisations, terrorists and individuals that act on behalf of their own intentions and interests – or that of a client.

These activities can be determined by economic, political and ideological causes. New kinds of infological weapons, such as malicious code, are constantly being developed. This also applies to methods for non-legitimate activities such as intrusions into computer networks or the dissemination and planting of viruses. Generally speaking, the means used are similar irrespective of the player category, while there are differences in the level of sophistication and complexity in the realisation as well as what kind of targets that are in focus for an assault.

Players and antagonists

There is a difference in definition of a player and an antagonist.[5] Players are individuals, groups and organisations that have the capacity to conduct malicious acts but are not doing so, due to lack of interest. Antagonists, on the other hand, are groups that have the motive, resources and capacity to carry out harmful operations against different kinds of targets.

---

5   Fylkner, M., Carlsen, H., Lewrentz, B. (2004) "Aktörer, antagonister och angrepp. En studie om det kvalificerade IT-hotet". Användarrapport. FOI.

Over time, a player can turn into an antagonist and vice versa. One example is the IRA (the Irish Republican Army); an organisation that – at least officially – says it wants to stop carrying out acts of terror. Whether it really has can be debated, it may just have changed focus for its activities – from terrorism to "ordinary" criminality. For instance, the IRA is suspected of being behind several bank robberies. Another example is left- and right-wing activists that use violence, or the threat of it, against political opponents. There is a possibility that they will escalate their activities and develop the desire and capacity to attack vital infrastructure and people who represent democratic values. Such attacks may be physical as well as virtual over the Internet. They have thus turned from players into antagonists. Apart from changes in attitude and motives, this kind of transition often requires greater resources, both financially and when it comes to skills.

The organisational structure of a player and/or antagonist differs, depending on the purpose of their mission, targets and available resources. Examples of diametrically opposed structures are hierarchical top-down systems, on the one hand, and networks put together ad hoc for a specific task and then disbanded after completion of the operation, on the other. The first kind may be an individual nation's intelligence and security service; the second, a group of cyber criminals acting independently and attacking specific targets. The more loose groups often use social media for contacts and co-ordination between the parties involved. In its structure al-Qaeda, for example, can be seen as a relatively independent system with autonomous cells in the top level management. There are also players that do not have, and do not want, any relations to other groups and instead act in solitary. The purpose is often security, minimising the risk of being discovered by staying under the radar of law enforcement efforts. One example of this is the Norwegian mass murderer Anders Behring Breivik.

Decisive factors for the way antagonists act are their driving force, their capacity to hit a target in a specific situation and their level of risk taking. Naturally, it is also influenced by what ethical boundaries and restrictions the player has. Before any kind of malicious activity, there is probably also some sorts of "economic" calculation or rational assessment as to whether the operation will succeed or not.

Players and antagonists in cyberspace can be divided into different categories, from fairly harmless but tiresome to much more dangerous, in some cases lethal.

A relatively unsophisticated category of players, whose behaviour often makes them easy to track on the Internet, are the *script kiddies*. Basically, these are youngsters interested in computers and networks that have downloaded information and software from the Internet on how to compromise accounts and files or initiate attacks on computer systems. In general, they lack the skills and knowledge to write more advanced computer programs. Their methods are based on scanning the Internet for information and trying to find any vulnerabilities to abuse. They may act individually or in groups.

The *cracker*, on the other hand, is a person who tries to penetrate security systems. It is the cracker who opens systems for attacks. The common method is software cracking, i.e. modifying the binary code which is the basic foundation of the software and disbanding any security restrictions. By doing so the application can be copied. A cracker can also be a hacker.

A *hacker* is a software programmer with very good skills in computing and the capacity to sneak into relatively advanced computer systems. These kinds of operations may have serious consequences for the victims. The purpose may be to steal passwords, look for information or manipulate and modify both hardware and software. It may be an act of economic criminality or part of bigger digital attack.

Hackers conduct their operations alone or in groups. Examples of hacker attacks include *defacements* of websites where information, pictures and texts are distorted or replaced. Through *distributed denial of service* (DDoS) attacks or spamming it is possible to saturate an opponent's servers and computers. This results in communication to and from the computers being made more difficult or impossible. In order to steal, manipulate or destroy information, hackers use "tool kits" with malicious code such as viruses, worms and trojans.

Hackers can be divided into white, grey and black (white hat hackers, grey hat hackers, black hat hackers) with variations. A white hacker follows an ethical code and tries to protect systems by pointing out flaws in them for the owner and administrators. Black hackers try to exploit the systems in illegitimate ways. Grey hackers are in between and act in both fields, sometimes legally, often illegally.

A *hacktivist* is a hacker and Internet activist with a political, ideological, religious or national cause, or combinations of these. A hacktivist may be a person who uses the net as a bulletin board to spread information and messages. In more serious cases, hacktivists conduct digital attacks against persons and companies with views that they don't share or like.

When it comes to *cyber terrorists* one can distinguish between "pure" cyber terrorists and "traditional" terrorists who use the Internet for co-ordination and spreading information. Pure cyber terrorists try not to be discovered but want to hide their intentions, at least until full scale digital attacks have been carried out. Cyber terrorism is in its nature quite safe, profitable and hard to discover. So far, there has not been a single (official) case of cyber terrorist attacks with deadly outcome. Attempts have been made with various degrees of success.

More or less all terrorist organisations – irrespective of their situation, geographical location and cause – use the Internet to look for information, co-ordinate resources and conduct operations of different kinds. One advantage for this kind of antagonist is modern society's dependence on working IT systems. Various sorts of vulnerabilities that are built into the systems are attractive targets for attacks. The Internet should not be regarded as a target in itself, but the control and management functions in critical infrastructure, "SCADA systems" that use the Internet, may be targets.

The basic logic of *cyber criminality* is money. The Internet is a formidable forum for all sorts of fraud, such as virtual robberies and blackmail, identity theft etc. Cyber criminality is directed against individuals, organisations and financial institutions. It is a truly global movement and an international phenomenon, not restricted by borders, geography, origin or national laws. It involves both individuals and large groups. A growing phenomenon is cyber mafias. These kinds of organisations work as conglomerates with major resources, and they are always interested in finding new lucrative areas and targets to exploit. Depending on the requirements, hired hands are brought in. There are presently a number of infamous groups on the net whose activities have a turnover of several billion U.S. dollars per year. There are suspicions of connections between cyber criminals and organised crime. These kinds of loose organisations will probably increase in the future.

One category that potentially could do serious harm is the *insider* in an organisation. He or she may be a person who acts arbitrarily and independently, without support from the outside. The driving force and motivation is often dissatisfaction with the work situation. Harmful activities can be initiated by a dismissal or a perceived injustice. Insiders can also be planted for a specific purpose, for instance people with good IT skills who sympathise with terrorists. Insiders can also be people linked to foreign intelligence services, a criminal organisation or a rival company.

One phenomenon that is steadily increasing is players acting as third parties. The common term for this is *cyber war by proxies*. These may be individual hackers or groups of hackers with very good IT skills. Behind such groups there are stronger parties, such as foreign intelligence services. This means they have access to very large resources.

The point of using third parties is to conceal the originator, motives and intentions behind an operation. Digital tracks that point to a specific party can be hidden in the cyber mist. Using third party activists is a well-working strategy, not only for states that want to avoid drawing attention to its activities on the Internet; the practice is also interesting to individuals and companies that are trying to get advantages over their opponents and competitors. Through cyber espionage via proxies, for instance, it is possible to steal sensitive and vital information. Using third parties, specific organisations and persons can be discredited by planting malign rumours and information. These actions may be part of a psychological operation with the purpose of influencing people.

In order to protect systems that are critical to society, such as military and security structures, many high-technological nations establish specific *information warfare units*. These units handle the whole spectrum of operations on the Internet and in the air, for offensive and defensive purposes. This kind of activity involves interception of information, disruption and destruction of the opponent's information and communications systems as well as deception and psychological operations.

Resources for cyber antagonism

The kinds operations that can be carried out depend on the amount of resources, the motives and driving forces of the players and the general security situation. An individual normally has less resources than groups and large organisations. Security and intelligence services usually have a lot of means at their disposal.

The resources can be of different kinds and vary over time. They may be personnel, finances, logistics and skills, they may involve access to IT, electronic warfare means and networks. Examples of infological weapons are logical bombs, viruses and trojans. Means for electronic warfare include equipment for interception and interruption of cellular networks, as well as microwave weapons that knock out electronics and components.

The lower the cost of conducting advanced cyber-attacks, the more likely they are to take place, as long as there are motives and driving forces for the players. The cost pressure applies both to purchases of "weapons", and the development of know-how and methods for illegal actions. Handbooks and instructions on how to construct bombs, intercept radio traffic and adapt systems in different ways can be downloaded from the Internet. Nowadays, advanced equipment for wire-tapping and jamming can be bought for couple of hundred dollars. If a potential antagonist develops the right skills, he can build the equipment himself. During the Cold War this was very difficult. In practise, only advanced intelligence and security organisations had the resources for this.

The openness of the Internet when it comes to the dissemination of information creates new challenges for law enforcement agencies. The general pressure on prices in combination with increasing access to information leads to gradually lowered barriers for malicious operations. A motivated and skilled antagonist can easily do a lot of harm to vital infrastructure with small means.

Procedures

Cyber antagonists act at a distance. An operation can be staged on one side of the world but have effects on the other. No costly apparatus is required. Somewhat incisively, all that is needed to conduct computer operations is a good computer and access to a broadband network. In contrast to kinetic attacks, for instance a bomb dropped on a target, it is hard to foresee the consequences of an infological assault. There is a risk of cascade effects and unintentional dissemination that affect third parties. There is no Geneva Convention that defines what can and cannot be done in cyber space.

One obstacle for potential antagonists is that activities in cyberspace leave electronic traces that can be followed, if IT forensics know what kind of tracks to look for. On the other hand, there are several methods that can be used to hide intentions. One effective method that is popular among more advanced antagonists is peer-to-peer networks[6] where communications and exchange of information are conducted in closed forums with invited parties. Security is reached by all parties involved guaranteeing each other's participation via a control system with continuous acceptance routines in the closed groups. These kinds of networks conduct operations ad hoc when needed, after which they are shut down, when communications and exchange of information have been made. The fact that the closed networks are used only for a short period of time makes it difficult for law

---

6   In a peer-to-peer network terminals are connected directly to each other in closed channels.

enforcement agencies to track and follow the operations of these individuals and groups.

Sensitive information can also be camouflaged. In the physical world, stealth technology is e.g. used. In the virtual world sensitive data can be hidden in streams of information. One method that is allegedly used by terrorist organisations is *steganography*. This means that information and important messages are embedded in software program lines at specific web sites. Antagonists that know of such websites visit them and read the messages. How common this is, if it is done and on what scale is hard to say. Using cryptology and "tunnelling" information between sender and receiver is a common way of acting among more skilled people involved in illegal activities. Another method in order to make detection more difficult is using false IP addresses when communicating.

One way of reducing the risk of interception of mobile communications is to use cell phones only once and then get rid of it. The large amount of stolen phones that turn up in other parts of the world, as opposed to where they were stolen, indicates that this is actually happening.

Threats

What kind of targets that might be of interest to cyber antagonists naturally depends on what they want to achieve and what resources they have. Assaults can be directed against symbolic targets, people and organisations, or in order to disrupt systems critical to society such as banks, financial institutes, telecommunications and computer networks, electrical grids, transmission links etc. The number of targets that can be attacked is huge, almost endless, which makes them very hard to protect. There are always different kinds of vulnerabilities, technological and human, that can be exploited by clever players.

In high-technological societies most activities, regardless of the line of business and concentration, depend on working systems for electronic communications. The systems are interrelated and interacting. A fault in one system can spread to other systems in a very short time. This implies the need to protect and secure functionality at all levels of the chain of command.

Electronic threats can be either physical or infological, of a direct or indirect kind. Physical IT threats include influencing hardware such as cables, wires and computer equipment by physical force, i.e. shutting down transformers or cutting cables. Infological IT threats include penetration of vital support systems, planting malicious code and distributed denial of service attacks. Such infological IT threats are conducted through computer and network operations, what is normally called cyber warfare.

Means and methods

Cyber warfare can be divided into Computer Network Exploitation (CNE), Computer Network Attacks (CNA) and Computer Network Defence (CND). The latter involves protecting systems, networks and software from illegitimate activities.

The purpose of exploitation is to get a picture of an adversary's IT systems, networks and equipment, what kind of procedures and conducts that are used. The objective is to locate and identify a specific terminal or node in a network or chart parts of or the whole network when it comes to structure and capacity. A party planning an attack tries to figure out how the opponent's hardware and software is designed by testing the system's reactions to different kinds of stimuli.

With hardware from well-known suppliers it is possible to find out any security obstacles and built-in weaknesses, for instance through *reversed engineering*. This means trying to understand the design of a device or application by taking it apart, if necessary breaking it. When it comes to software, a *disassembler* can be used in order to study the program code at very low levels. Through experience and detailed knowledge about what kind of architecture is used, an attacker can redefine the source code of the program, making it possible to understand if there are any backdoors or bugs in it. Operating systems such as Windows and many software programs have weak points and embedded faults that can be exploited.

Fingerprinting is both a method to identify what kind of programs or operating system that are used by analysing the system's packet data, and a way of analysing the packet data system. If the hardware and software of the opponent/victim are known by the attacker, it is enough to study database information on certain web sites that describe security holes in different products.

Once the exploitation has been made the barriers to conduct some kind of digital attack is low. The step from penetration of a system to a full scale assault is very short.

Infological attacks can involve a number of things, from surveillance of the adversary's traffic to interception and destruction of his equipment. Disruptions are graded based on how long it takes to restore the system, from a couple of hours to several days.[7] An attack can either be *active* – meaning that data, files and information are modified – or *passive* – interception or analysis of data traffic between two or several parties in a network by a third party. Passive attacks are very hard to detect, since they neither destroy nor distort data.[8]

Examples of different kinds of IT attacks are breaches, planting malicious code, distributed denial of service attacks and bugging Internet traffic with the aid of sniffers.

The way a breach is performed depends on the perpetrator's skills and resources. Port scanning, i.e. a qualified hacker identifying what kinds of ports are open and available for attacks, is often the first step.[9] The next is to break into the actual system. Logging information and passwords is often required.

One problem for hackers is that passwords may be enciphered. However, often there is no need to crack the password at all; it is possible to get into the system anyway through *social engineering*. This means that the hacker contacts the system owner or the user and asks for the user name and password in a polite and convincing way. If this method is not available, it is always possible to use personally developed or downloaded software.[10] There are good programs available for this purpose; a hacker can download poorly protected (wrongly configured) password files over a system and quietly crack them at home.

One difficulty for hackers is that penetrations can leave digital traces in the log files of the attacked system.[11] Therefore it is important for the hacker to hide all traces left behind after such a breach. One way to reduce

---

7   Three levels are used: *disruption* – the system is shut down for 4 hours, *denial* – the system is shut down from 4 to 72 hours and *destruction* – it takes at least 3 days to recover.
8   Stallin, W. (2003) "Network Security Essentials. Applications and Standards". *Pearson Education, Inc.*, New Jersey, USA. 2003 (2nd edition).
9   In many cases port scanning is not necessary, but if the objective is to attack web servers, port 80 is normally open for intrusion, just as port 21 is open on ftp servers. Port scanning is a viable method when the system that is going to be approached is unknown. It is possible to download software for this purpose from the internet.
10 There is a risk that personally developed programs are easier to discover than downloaded software. This, of course, depends on the skill of the hacker.
11 It is not always certain that the breach is logged; it depends on the system settings.

the risk of discovery is to plant misleading information about who is behind the operation. The hacker may use false identities or script kiddies, i.e. a third party, as cover.

Penetrations can also be performed through someone having implemented back door functions in the system in advance; code hidden in circuits or software that opens up for attacks. This can be done by insiders or be planted into the system before purchase. Malware[12] is a common name for malicious, more or less autonomous, software such as viruses, worms, trojans, adware or key loggers.[13]

According to the Swedish Standardisation Institute (SIS) malicious code is "... a code which, when executed, deliberately causes disruption or harm to the activity".

Malware can be divided into *reproducing* and *non-reproducing* versions.

Viruses are an example of reproducing malicious code. The primary target for viruses is to self-copy and spread to other uninfected parts of a system. A virus needs host program in order to reproduce itself. They can be transmitted via e-mail, ActiveX, scripts on websites, links to websites and jpg images. One example is Melissa, which infected Word 97 and 98. Apart from destroying stored information, this kind of virus can also send information to third parties.

There are different kinds of viruses.[14] One is *system viruses*, which create loss of memory through infections of the part of the system that starts the programs. Another kind is *execution viruses*, which seek for files ending in .exe or .com and infect them, at which files may be erased. Other kinds include *stealth viruses* and *encrypted viruses*, which hide that they have infected files, *polymorphic viruses* that change their code every time they have infected another program and *macro viruses*, which only infect certain programs.

Worms, as opposed to viruses, don't need a host in the form of a program or a file to reproduce; instead they copy themselves via LAN connections or on the Internet. They can e.g. spread through e-mail attachments. Worms often use security holes in operating systems. By auto-reproducing, the number of worms can increase more or less exponentially over time, which naturally puts a lot of pressure on the system. One effect of a worm

---

12 Source: Wikipedia.
13 A keylogger can also be hardware-based, e.g. attached to a keyboard flex, in order to intercept signals.
14 Bishop, M. (2003) "Computer Security Art and Science". *Pearson Education, Inc.*, New Jersey, USA. 2003.

attack is slower performance. Some examples of infamous worms are Nimda, Code Red, MS Blaster and Stuxnet, which among other things infected the Iranian nuclear program in 2011.

Trojans can be described as non-reproducing malicious code with the purpose of stealing information or destroying data. Trojans occupy and abuse system resources by secretly conducting services such as logging passwords and keyboard strikes; they can also borrow processing time. In contrast to viruses and worms, trojans are often under direct command of humans. They are transmitted through e-mails, sent by an antagonist, or implanted in links that can be downloaded from the net. They can hide in computer games, screen savers or other kinds of applications. By using Trojans, hackers can sneak into systems and take over computers.

Spyware are programs for espionage with the purpose of stealing passwords or logging the web browsing behaviour of users. This kind of program can be embedded in freeware that is downloaded from the Internet. The captured information is sent to the attacker. There are different kinds of spyware, for instance tracking cookies for information on what kind of web sites a user visits.

The purpose of *distributed denial of service attacks,* DDoS[15], is to saturate, or "sink", websites, by denying access to other users. Examples of targets for DDoS attacks include the websites of authorities and corporations. The aim of such attacks is to blackmail or undermine an opponent's capacity to spread information about his activities. The procedure is the following:

First the attacker penetrates a computer and installs a software program, a "bot"[16] with malicious code that opens up for take-over and control of the computer. The computer becomes a *zombie*.[17] After this, infected code, in the form of worms, is planted in e-mails or on websites, and disseminated to a large number of other computers, which in turn are compromised and create a botnet.[18] Usually, vulnerabilities in the abused operating system are exploited. When the attacker has set the time for a saturation attack, the infected software is activated and all zombie computers in the bot network

---

15 DDoS – Distributed Denial of Service.
16 A "bot" is an abbreviation of robot. Installation of the malicious code means the computer can be taken over, without the owner's knowledge, which opens up for remote monitoring and attacks.
17 A "zombie" is a computer infected by a bot.
18 A "botnet" is a huge number of bot-infected computers that are controlled by a third party.

start communicating with the website under attack. The charge that follows is so intense that the website crashes.

There are different ways of intercepting data traffic over the Internet. One way is to use "sniffers". A sniffer is usually software that can monitor and log traffic over a network. The method is based on "taking" data packets, deciphering and analysing the contents.[19] Depending on the network structure, traffic can be sniffed via the ports in a hub, switch or router. Some network hardware is prepared for sniffing via a *SPAN port*, where all traffic is copied when it passes though the other ports.

Sniffers can be used to analyse own network problems, discover penetration attempts and monitor computer usage. With the help of sniffers, statistics and traffic data can be produced and unwanted or suspicious content in the network traffic filtered

---

19 Normally a "sniffer" is not able to decipher enciphered information unless the cipher is poor or a large amount of computer processing power is used. On the other hand, sniffers can be used to intercept passwords and personal identification numbers. There are also "air cracks" and "packet sniffers" that can be used on wireless networks.

The Superpowers

Information warfare is by no means a new phenomenon in history; it is as old as mankind. Manipulation, deception, eavesdropping, saturation, withholding and destruction of information for a specific purpose and aim are part of human behaviour. Through the years, information has been used in different ways in military, socio-economic and political contexts in order to influence both opponents, own forces and compatriots. The introduction of information technology – computers, the Internet and mobile communications – opens up for the development of new means and methods both when it comes to influencing third parties and protecting information and information systems.

From a modern perspective, information warfare is based on a blend of ideas, theories and concepts, prompted by various security policy environments, situations and contexts. To a large extent it depends on the technological development. A number of thinkers and analysts – particularly from Russia and the United States, but also from China – have contributed with ideas. However, it is important to stress that there is no single universal theory defining information warfare; leading nations with ambitions develop their own ideas and concepts on what should be included in the term and how different resources can and should be used. All the superpowers mentioned attribute information warfare a growing importance and in some cases a decisive role in present and future conflicts. Every large military and political conflict contains some kind of cyber component.

Network centric warfare

One of the first persons to point out the paradigmatic shift towards digital warfare was the Soviet Chief of the General Staff, Marshal Nikolai Orgakov. In 1984 he introduced the term Military-Technological Revolution (MTR) to describe the fundamental change from mass armies to technology-driven operations. From Orgakov's point of view, the decisive factor to defeat the enemy is not the number of tanks or aircraft, but the quality and degree of high technology in them. Advanced precision weapons win over quantity. During the 20th century, two phenomena fundamentally changed warfare, Marshal Orgakov says. The introduction of airplanes, motorised vehicles and chemical weapons during the First World War, and nuclear weapons, computers and missiles during and after the Second World War.

The opportunity to combine different weapons systems – such as precision weapons, computers and sensors – created conditions for a fundamental shift. However, the Soviet Union lagged behind its rivals. The Soviet General Staff had problems co-ordinating different military branches and units. At brigade level, they created a combined intelligence and suppression system for the air defence. But the Soviet system was under a lot of pressure, and resources were limited. A national collapse was on its way. The breakthrough for the new thoughts did not come about.

However, the ideas were adopted by the Pentagon in the late 1980s, under the term *Revolution in Military Affairs* (RMA). Gradually the meaning of the term was expanded with new ideas and thoughts.[20] By that time the United States had become the leading nation in the development of modern warfare based on the concept of *Network Centric Warfare* (NCW).

A revolution in military affairs can be defined as a radical shift of common views and understandings that fundamentally changes perceptions and overall strategies. It may be the introduction of a new weapon, a new technology or new ways of combining and co-ordinating different systems to working units. A need for new structures arises in order to wage war with new means and methods. With reference to Watzlawick's theories, a revolution in military affairs can be viewed as a second order change.[21] This kind of radical shift usually occurs every thirty to forty years.[22]

A revolution in military affairs often has its breakthrough during war. This is when people realise that the old methods no longer work. In general, it is often the losing side that concludes it has to remodel its strategies, military doctrines and organisational structures. One historical example of a revolu-

---

20 Mowthorpe, M. (2005) "The Revolution in Military Affairs (RMA): The United States, Russian and Chinese Views". *University of Hull*, vol. 5, No. 2 (summer).
21 Watzlawick defines a first-order change as a change within a system, for example in terms of investment in new production capabilities and equipment etc., whereas a second-order change involves a fundamental change of the whole system, leading to a completely new way of working and logics. Watlzawick, P. (1976). "How real is real? Confusion, disinformation, communication". *Random House*, New York, USA.
22 "Dags för ett nytt RMA". FOI Framsyn nr 2, 2001.

tion in military affairs that changed the character of conflicts is the *Blitzkrieg* concept, defined by Germany in the years before the Second World War. By combining dive-bombers and tanks it was possible to co-ordinate strike force and mobility. This way, the enemy defence lines could be opened up. Germany had learned the lesson of entrenchment and mass slaughter from the Western front in World War I.

Another example is Napoleon's idea of mass armies. This is in its turn is the basis of the term total defence, meaning all of society should be organised for war. Everything from public schools to industries and workshops are cogs in the machinery. Conscripts play an important role.[23] The longbow, introduced by the British at the battle of Agincourt in 1415, was a revolution in military affairs that gave strength to poorly armed units against knights in full armament. The battle was decided at a distance, without making full use of the knights' advantage in close combat.[24]

From a security policy view it is very important to analyse technological, doctrinal and societal changes. In order to understand what kind of revolution in military affairs we are facing, it is vital to learn from previous ones and what consequences they had for warfare. This is the foundation for any military policy and security conclusions and decisions.

Despite its military and economic superiority, the United States was humiliated in the Vietnam War. It did not succeed in subjugating a much weaker and less developed nation, which was assisted by China and the Soviet Union, the ideological opponents of the United States. Almost 60,000 American soldiers and two million Vietnamese died during the conflict, which lasted more than ten years. As a consequence, both the public and military personnel questioned the U.S. strategies and actions. A few years later, in 1979, the failed attempt to liberate the American hostages at the embassy in Tehran increased the negative public opinion. The time was right to define new strategies and doctrines on what could be seen as specifically American national advantages, especially advanced technology.

---

23 During the Cold War the Swedish Armed Forces could mobilise up to 800,000 men.
24 In Agincourt in northern France, a decisive battle was fought between the British and the French in 1415. Due to a number of successful factors, especially the introduction of the long bow, the outcome of battle was in favour for the British, despite less armed and trained men. The British had 7,000 soldiers, of which 5,500 were bowmen, against a French force of 20,000 well equipped soldiers, including several thousand knights.

Examples of technologies that were developed and refined from the 1970s on include laser-guided bombs, stealth technology, autonomous systems (such as drones) and satellites, e.g. the global positioning system, GPS, which was first installed in 1973. The system is used both for location and time. This is a key technology for co-ordination of and co-operation between units and is the basis for what is called navigation warfare (NAVWAR). Information technology with super computers and smart sensors, Artificial Intelligence (AI) and the Arpanet, which later developed into today's Internet, are other key technologies.

In 1983 the Reagan administration introduced the *Strategic Defence Initiative* (SDI), or "Star Wars". The purpose was to demonstrate technological superiority. In propaganda films laser weapons were shown destroying Soviet satellites and incoming nuclear missiles.

In the Soviet Union, President Mikhail Gorbachev tried to reform society and the economy through *glasnost* and *perestroika*. The war in Afghanistan, started during Christmas 1979, coupled with an inefficient production apparatus, bureaucracy and corruption, drained the country's resources. The regime had severe difficulties keeping an even pace with the West, both technologically and financially. In some fields of science and technology the Russian capacity was still good, thanks to its education system and advanced research institutes. Despite attempts to create a more open and liberal economic climate in the Soviet Union and the other satellite states, the system gradually fell apart – from the inside and at an increasingly fast pace. The Berlin wall was toppled in 1989, and the Warsaw pact was disbanded in 1991.

Afterwards, it has emerged that Star Wars to some extent was a deception operation.[25] One of the intentions was to trick the Soviet Union to "arm itself to death" in its efforts to compete with the United States, despite the fact that its resources were stretched to the limit.[26] This, together with the American help to the Afghan *mujahedeen* had decisive significance for the speed of the breakdown.

After the collapse, the Russian military sought a new role, purpose and task. The Gulf War in 1990–1991 demonstrates a radical shift in military

---

25 Beichmann, A. (1993) "Reagan ought to get Oscar for Star Wars – how former President Ronald Reagan used the threat of proposed Strategic Defense Initiative to weaken the Soviet Union". Column. *News World Communications, Inc.*
26 Graham, J. (2000) "Military Power vs Economic Power in History".

strategy and conduct. Based on theories of command and control warfare,[27] the American-dominated coalition force attacked strategic Iraqi points – such as military staffs, radar stations and networks – with laser guided precision weapons. Vital parts of the Iraqi communications and information infrastructure were knocked out.

Command and control warfare is by no means a new concept; the inspiration comes from 19th century military theorist and thinker Carl von Clausewitz. According to von Clausewitz, both enemies and allies have a number of centres of gravity (*schweerpunkte*). If they are knocked out – kinetically through bombardment, or digitally through disruption of radio and radar systems, computers and networks – this may affect parts that are decisive for the whole system. Without the possibility to communicate, the system will gradually collapse. During the Second World War both sides used strategic bombing against vital enemy facilities and infrastructure, but precision was not good enough. It was not until laser-guided missiles and smart bombs were introduced that the technology became advanced enough to fulfil the precision requirements. But only a small part of the bombs used in the Gulf War were "smart", the greater part was conventional.

Since the Iraqi military system was hierarchical and top-down, the effects of the precision bombings were severe. When the Iraqi chain of command broke down, the system disintegrated – very fast. Less than a hundred hours after the first allied tanks crossed the border into Kuwait, the military part of the war ended. The military objective had been reached (although not necessarily the political objective).

The Russians, on their hand, have identified a key factor behind the victory of the U.S.-led coalition; *information superiority*. The party that has access to information – and can co-ordinate it strategically, operationally and tactically – will get a decisive advantage over its enemy. If one side can deprive the other of access to information, it will affect the latter's capacity to make good decisions.

The result of the conflict came as a shock to the Russian armed forces, since the Iraqi arsenal to large extent was Soviet. Saddam Hussein's army was modern and one of the biggest in the region. The fact that vital parts could be knocked out – in a very short time – through command and control warfare was an alarm bell, not just to Russia but also to China. Doctrines based on Cold War structures had been proved antiquated. The era of mass armies was definitely over. A new way of thinking was required.

---

27 Command & Control Warfare: C2W.

Marshal Orgakov's ideas on a military technological revolution were brought out, and Russian analysts suggested that a new paradigm, based on information and networks, was in the offing.

As a result of the rapid development in information technology and the experiences from the Gulf War, the United States adopted a new military concept called *network centric warfare*. During the 1990s the U.S. became the leading nation in high-tech warfare, based on network centric logic.[28] The background to the concept can be found in the problems that the U.S. Navy faced with Soviet threats against American carrier battle groups in the 1980s.[29] One problem was the difficulties for the group to discover enemy forces at large distances, another to control its own weapons systems, such as fighter aircraft and Tomahawk missiles. As computers, including those on the smaller vessels in the groups, became more powerful and advanced, information could be processed and forwarded to sensors and data bases on the other ships in the groups. This meant that the units in the carrier group could be connected to bigger networks. This in its turn led to increased efficiency and better co-ordination of command and control.

In 1996 Admiral William Owens published a report discussing some of the basic elements in network centric warfare. He stressed that the development of intelligent sensors, command and control systems and precision weapons increase efficiency and the capacity to create immediate situational awareness.[30] The same year the Joint Chiefs of Staff published *Joint Vision 2010*, a report that described how the American armed forces were going to gain information superiority[31] during operations.[32] The first time the term "network centric warfare" was mentioned more explicitly was in an article published in a U.S. Navy journal in 1998, written by Vice Admiral Arthur K. Cebrowski and John Gartska. The concept was elaborated in a book entitled *Network Centric Warfare*, published in 1999.[33] During the

---

28 Mowthorpe, M. (2005).
29 "Nätverksbaserat Försvar. Fördjupade studier utifrån Försvarshögskolans sakområden". (2003). *Swedish National Defence College*. Stockholm, Sweden.
30 Owens, W. (1996) "The Emerging U.S. System-of-Systems". *National Defense University*, Washington DC, USA.
31 In American and British nomenclature the term DBA – Dominant Battlefield Awareness and Information Dominance – is used.
32 Http://www.dtic.mil/jv2010/jv2010.pdf.
33 Alberts, D., Gartska, J., Stein, F. (1999) "Network Centric Warfare: Developing and leveraging information superiority". *CCRP Publication Series*. Revised August 1999 (2nd edition).

first years of the 21st century a number of books and research papers on the topic have been published.

The basic philosophy in network centric warfare is based on systems theory, which views military structures as a big network consisting of staffs and command centres, units, sensors and weapons systems that are physically and logically linked to one another. Platforms such as aircraft, ships, tanks and so on are sensors connected to the all-embracing network and other sensors. Within the platforms there are a number of smaller networks consisting of personnel and technical subsystems. Through the network it is possible to optimise resources after one's own needs against the opponent's capacity. The network is used as a common resource for co-ordination and concentration. The goal is for the total effect to be bigger than the sum of the individual actions.

The basis in the network is information. By collecting information from all parts and sensors of the network, known as "information fusion", and distributing it more or less in real time to the units that need the information, a joint situational awareness is created. This means that the network and its users act like a living organism, which perceives and understands a situation as a whole; that way it can act with the right means at the right level of the chain of command against different kinds of events and situations. In other words, the network becomes *self-organised*.[34]

A vital component is information superiority, i.e. denying the enemy access to information and information systems that are used to define situational awareness. Without information and the opportunity to communicate it is hard to get an accurate picture of the chain of events. The capacity to act is reduced.[35]

The parties that have to communicate with one another do so directly, without going through more senior levels of command. This implies an effective distribution of information in more or less real time to the units that need it. One consequence is that the hierarchical military system will be broken down and the organisation becomes flatter. The phenomenon is called "Power to the Edge" in a book by the same name.[36] In order to realise the idea, the writers stress the importance of creating a global network of interconnected networks of computers and sensors, a *Global Information Grid* – GIG.

---

34 Alberts, D., et al (1999).
35 Alberts, D., et al (1999).
36 Alberts, D., Hayes, R.E. (2003) "Power to the Edge. Command, Control in the Information Age". DoD Command and Control Research Program. CCRP Publication Series.

The vision at strategic level – general staffs and military headquarters – is for the network to give an overview of the different events in a conflict. Based on this it will be possible to manage and act directly at operational and tactical levels. The purpose is to achieve transparency at every stage. Through network logic, what Clausewitz called *the fog of war* will be overcome.[37]

The theory of network centric warfare is one of the corner stones in the transformation of the U.S. armed forces, and the Department of Defence maintains that the GIG is a fundamental part of it. A number of countries have adapted the theories. Great Britain, for instance, is developing its own version of NCW, designated Network Enabled Capability.

While the network centric ideas are being established and realised – such as the process of connecting different kinds of networks to an enormous GIG – criticism against parts of the concept is growing. This applies especially to the lack of discussions on the risks of information warfare through deception, wire-tapping, jamming and destruction of networks and the information that is processed, distributed and used in the systems. Other challenges that the critics claim are disregarded include the need for robust networks, the issue of how to distribute bandwidth to the units and platforms that are part of the networks, the risk of illicit tapping of information, such as cyber espionage, and how to handle vital and secret information in a network accessible to lots of people.

Based on network and command and control warfare, the United States is developing doctrines for information warfare. The first time the term was mentioned officially by the Department of Defence was in a framework document in 1992, describing the development of concepts for command and control warfare. During the end of the 1990s the different service branches established their specific view on what should be included in the term. Especially the U.S. Air Force was a driving force, partly because it is seen as perhaps the most technically advanced service branch, partly because it wants to control the issue. Parts of the U.S. Army were not altogether pleased about USAF's ambitions. The U.S. Marine Corps and the U.S. Navy developed their own definitions and concepts, adapted to their specific activities.

At the Department of Defence there was a need to co-ordinate the different views. A number of documents were produced, such as the joint

---

37   Clausewitz, C. von (1832) "Om kriget".

service branch document *U.S. Joint Pub 3-13 Information Operations* of 2006[38] and the visionary *Joint Vision 2010* (JV-2010) and *Joint Vision 2020* (JV-2020). The term information warfare was also replaced with the less aggressive *information operations* – IO. Every branch established its own specific IO resources and units.

In the *Information Operations* doctrine of 2006, IO is defined as consisting of five basic capacities; electronic warfare (EW), computer and network operations (CNO), psychological operations (PSYOPS), military deception (MILDEC) and operations security (OPSEC).[39] The purpose of IO is, according to the doctrines, to influence, disrupt, corrupt or create unpredictable behaviour in or otherwise influence the enemy's human and automated decision-making while protecting one's own resources. Information operations can be used both offensively and defensively. In the autumn of 2010 the doctrine was modified.

Electronic warfare involves signals and communications intelligence against radar and radio systems, including positioning and identifying enemy equipment, monitoring radio signals and different kinds of jamming, blocking and saturation of enemy systems. The purpose is to deny the enemy access to his own information resources. Sensors are misled through false signals; this affects situational awareness and the capacity to make the right decisions at senior levels of command. Technologies such as lasers can be used to blind cameras and frost lenses. Electronic components used in airplanes and vehicles can be knocked out by electro-magnetic bursts.

Computer and network operations on the Internet, e.g. what is normally known as cyber warfare, can be divided into different categories. The opponent's systems, networks and computers are identified through exploitation. The information is used to find vulnerabilities that can be exploited for intrusions. The purpose of intrusions may be to steal, manipulate and/or destroy information and information systems. This is done with the aid of malicious code such as viruses, worms and trojans. Through Distributed Denial of Service attacks, DDoS[40], it is possible to saturate – spam – the enemy's networks and computers, so that they stop working and the system is shut down as long as the operation lasts.

Psychological operations aim at influencing human perceptions and behaviours. The objective is to steer the enemy in a direction that is beneficial

---

38 US Joint Pub 3-13. "Information Operations", 13 February 2006 (updated version).
39 US Joint Pub 3-13. "Information Operations", 13 February 2006 (updated version).
40 DDoS – Distributed Denial of Service.

to one's purposes. Examples of methods include deception and campaigns using mass media such as radio, TV and newspapers, but also social media on the Internet. Both electronic warfare and computer and network operations are important parts of psyops, e.g. in order to undermine an enemy's trust in his own systems and the information they supply.

Military deception occurs in more or less all forms during information operations, through different methods and instruments. Operational security means protecting one's own operations, units and systems from hostile information operations.

During the 21st century the number of cyber attacks against vital U.S. information infrastructure has accelerated; it now constitutes one of the most serious strategic threats against the nation. The high technological superiority based on network centric logic has a flip-side and an Achilles' heel; the great dependence on a working IT infrastructure.

The network approach (to connect smaller networks into bigger units), the dependence on time and location co-ordinates from the GPS system, the focus on more powerful processors and computers; all this has led the United States to become vulnerable to hostile information attacks, especially in the form of cyber warfare and espionage over the Internet. In different forums and contexts, the United States has singled out China as the greatest threat in the cyber arena, along with Russia. In order to protect its sovereignty in the digital sphere a U.S. Cyber Command was established; it became operational in the autumn of 2010. In May 2011 a new doctrine was launched, according to which the U.S. reserves the right to launch a counterattack against any major cyber operations against vital information infrastructure. Moreover, the United States encourages NATO to prevent cyber attacks; any major attacks against one of the countries in the alliance shall be seen as an attack on them all. In January 2012, the Obama administration stipulated that three areas are especially important for national security: special operations, autonomous systems, such as drones, and cyber warfare. A new era of warfare is in the making. Clausewitz's ideas on the fog of war are still relevant – especially in cyberspace.

The Russian view on information warfare and information operations

The development of doctrines and strategies in a country must be understood in and related to a wider context. It is based on a whole range of factors and assumptions such as historical experiences, geographical tensions, different military threats, economic situation, ideological background, technological standards, and the country's constitutional foundation – for instance, what kind of leading players and institutions it has. Russia is by no means an exception. With the end of the Cold War and the transition through the years of instability to a society based on strong leadership, the "strong state" of President Vladimir Putin has influenced ways of thinking. For obvious reasons, the modern Russian experience differs from the West's. This affects its military thinking in general and more specifically its views on information warfare.

There are presently no open, official Russian doctrines or policy documents specifically describing information operations and information warfare that would correspond to the U.S. Joint Publications. However, there are several doctrines and strategic documents that describe threats to Russia's national sovereignty. These include the Military Doctrine[41] and the Doctrine on Information Security of the Russian Federation[42], both from 2000 and approved by the Security Council. Three kinds of military conflicts are identified as threats to Russia.[43] The first is an escalating conflict in the immediate surroundings of Russia's borders. The second is a direct confrontation with the United States and its Western allies. The third is a possible conflict with an expansive China. The risk of the two latter is assessed to be low.

In the Information Security Doctrine, the security policy discourse extends into the domain of information.[44] The document discusses information-

---

41 *Voyennaia Doktrina Rossiiskoy Federatsii. Utverzhdena Ukazom Prezidenta RF ot 21 aprelya 2000 g. No. 706.* Referenced in Sokov, N. (2004) "Russia's 2000 Military Doctrine".
42 *Doktrina Informatsionnoi Bezopasnosti Rossiiskoi Federatsii.* The document is translated and discussed in Carman, D. (2002) "Translation and Analysis of the Doctrine of Information Security of the Russian Federation: Mass media and the politics of identity". *Pacific Rim Law & Policy Journal Association.*
43 Leijonhielm, J., Hedenskog, J., Knoph, T., Larsson, R., Oldberg, I., Roffey, R., Tisell, M.,Westerlund, F. (2008) "Rysk militär förmåga i ett tioårsperspektiv – ambitioner och Utmaningar". Användarrapport FOI.
44 Carman, D. (2002). "Translation and Analysis".

related threats to Russia and how the state should act in order to guarantee the protection of strategically important information. The doctrine should be viewed as a policy instrument primarily focused on Russian society, but it is also intended to influence an international audience.

Internationally Russia has also been active, proposing its view on information security. Together with China, Tadzhikistan and Uzbekistan it signed an agreement in the autumn of 2011 defining a "common code of conduct" – the "Yekaterinburg convention". One purpose was to promote proper behaviour in the cyber arena.

As opposed to the Western approach, Russian analysts emphasise information-psychological processes in terms of protecting their own society from the influence of information put out by an adversary.[45] The struggle for control covers not only the concern for state security, but also the stability of the regime, including the personal interests and ambitions of leading players. Different interests among different players come together in a common view on the need for state security.

The Russian view should be understood in the perspective of the disintegration of the Soviet Union. Several analysts argue that one of the major reasons for the break-up of the Soviet Union was enemy psychological operations.[46] The feeling of vulnerability to foreign campaigns and the impact these operations had on society were subjects of intense discussions in the aftermath of the Cold War.

Other experiences – such as the Gulf War in 1990–91, the war in Afghanistan in 1979–89 and the wars in Chechnya in 1994–96 and the one started in 1999 – have also influenced the Russian way of thinking. From a psychological warfare point of view Russia suffered severe problems in Afghanistan and failed to exert an influence over its adversaries.[47] The conflicts in Chechnya also provided practical knowledge of and insights into the approach to information warfare.

---

45 Thomas, T. (1998a) "Dialectical versus Empirical Thinking: Ten key elements of Russian understanding of information operations". FMSO Special Study Center For Army Lesson Learned. Fort Leavenworth, KS 66027–1327.
46 Thomas, T. (1998b) "Russia's Information Warfare Structure: Understanding the roles of the Security Council, FAPSI, the State Technical Commission and the military". *European Security*, vol. 7, No. 1 (spring). Hoffman, D. (2008) "KGB Comes in from the Cold". *Washington Post*, 8 December 2008. Quoted in Carman, D. (2002) *Translation and Analysis*.
47 Serookiy, Y. (2004) "Psychological-Information Warfare: Lessons of Afghanistan", *Military Thought*, vol. 13 No. 1.

Both wars in Chechnya showed that in some areas even a small and relatively impoverished enemy can achieve information dominance over a stronger opponent by using the media component efficiently. The Chechens were much more flexible than the Russians in using the Internet and other tools to broadcast their perspective and gain influence over public opinion.[48] This was evident to the Russian military after the first war in Chechnya.

One of the main components of the Information Security Doctrine adopted by the Security Council in 2000 is to guarantee the protection of what is called "strategically important" information from foreign activities directed against the interests of the Russian Federation in the information sector. The doctrine is a synthesis of the official position and the state policy for maintaining information security. It has been interpreted as the ultimate authority of a nation state to regulate its information and media networks, for instance, by nationalising free media.[49] The doctrine discusses a wide variety of issues – not only the need for protection of networks and information but also how to strengthen national identity and preserve the cultural heritage in order to ensure that the young generations develop constructive moral values, patriotism and civic responsibility, for the well-being of the country.

The Information Security Doctrine has been a valuable tool for the Kremlin to get a grip on the flow of information in Russia. By nationalising media such as NTV and other free channels, the state has created an instrument for monopolizing the truth.[50]

In the spring of 2010, a new Russian Military Doctrine was published. In it Russia stresses the importance of information warfare during the initial phase of a conflict in order to weaken the command and control capacity of the enemy. The Russians have learned from the American experience of command and control warfare. Subsequently, it is important to create a positive picture for the international community, in the form of an information campaign during the actual battle.[51]

Although there is no defined and officially sanctioned doctrine on information warfare in Russia, there are of course a good many theories and concepts presented by leading scientists, analysts and military specialists. The

---

48 Thomas, T. (2003) "Manipulating the Mass Consciousness: Russian & Chechen information war. Tactics in the second Chechen–Russian conflict". 14 April 2003.
49 Carman, D. (2002).
50 Carman, D. (2002).
51 Vendil Pallin, C., Westerlund, F. (2010) "Russia's Military Doctrine – Expected News". *RUFS Briefing* No. 3, February. FOI

Russians' views have been discussed publicly in scientific papers and conference presentations since the mid-1990s. They are seen as having some bearing on today's situation. The theories have been influenced by the debate on the Revolution in Military Affairs (RMA) and the Military Technological Revolution (MTR), as well as the concept of command and control warfare and network centric warfare.

The American analyst Colonel Timothy Thomas points out that there are several unique elements in Russia's approach to information warfare. Due to lack of resources and budget constraints during the 1990s, in the aftermath of the Cold War, Russian scientists spent more time on IO theory than the West. (The West focused on practice over theory.) But this could be to their advantage in the future, since there is nothing as practical as good theories. There is also an opportunity to learn from mistakes made by more advanced competitors. The Russians have developed some unique elements in their view on and use of information warfare.[52] Additionally, Russia has a historically very good education system with cutting edge research institutes in fields such cryptology, smart algorithms and advanced computer programming.

Several leading analysts claim that information technology will be a formidable weapon in the 21st century, fully comparable to weapons of mass destruction.[53] The importance of information warfare and its consequences was stressed by the Russian Chief of the General Staff, Viktor Samsonov, in the spring of 1996:

> The effectiveness of "information warfare" systems in combination with precision weapons and "non-military means of influence" makes it possible to disorganise the system of state administration, hit strategically important installations and groups of forces, and affect the mentality and moral spirit of the population. In other words, the effect of using these means is comparable with the damage resulting from the effect of weapons of mass destruction.[54]

---

52 Thomas. T. (1996) "Deterring Information Warfare: A new strategic challenge". *IWS – the Information Warfare Site*. 7 November 1996. Thomas, T. (1998a). Thomas, T. (1998b).
53 Fitzgerald, M. (1994) "Russian Views on Electronic Warfare. The growing role of information technology is rapidly lowering the barrier between war and peace". PowerPoint. http://www.nationalstrategies.com. Korotchenko, Y. and Plotnikov, N. (1994) "Information is Also a Weapon: About what should not be forgotten when working with personnel". *Krasnaya Zvezda*, 17 February 1994.
54 The statement by General Viktor Samsonov is the following: "The high effectiveness of 'information warfare' systems in combination with highly accurate weapons and 'non-military means of influence' makes it possible to disorganize the system of state administration, hit strategically important installations and groups of forces, and affect the mentality

Comparing information warfare to weapons of mass destruction is interesting and somewhat disquieting. At a Moscow conference on information warfare in 1995, one analyst, V.I Symbal, summarised several high ranking officers' views on cyber war as the second most serious, after nuclear war:[55]

> ... from a military point, the view on Information Warfare against Russia or its armed forces will categorically not be considered a non-military phase of a conflict whether there will be causalities or not [...] considering the possible catastrophic use of information warfare means by an enemy, whether on economic or state command and control systems, or on the combat potential of the armed forces [...] Russia retains the right to use nuclear weapons first against the means and forces of information warfare, and then against the aggressor state itself.[56]

The 2001 Congress report on cyber warfare states that Russian cyber warfare activities have a military role in the sense that gaining and holding information advantage over an enemy are substantial goals.[57] They could be accomplished by using specific information capacities to affect an enemy's information systems, decision-making processes, command and control systems, even the population.[58] Viruses and other information-related weapons could be used as force multipliers.

In coming conflicts there will be no clear battle lines and the fighting will take place in several dimensions and arenas, including space. Warfare has shifted from being a duel between weapons systems to being a duel between information systems. The arms race is moving into the digital sphere of software and algorithms.[59] Those with the most rapid and advanced computer capacities will gain an advantage over their adversaries.

---

and moral spirit of the population. In other words, the effect of using these means is comparable with the damage resulting from the effect of weapons of mass destruction."
55 Tsymbal, V.I. (1995) "Kontseptsiya Informatsionnoi Voiny" (Concepts of Information Warfare). A presentation on the Russian–U.S. conference on Evolving Post Cold War National Security Issues, Moscow, 12–14 September, quoted in Thomas, T. (1996) "Russian Views on Information-Based Warfare". The text is published in *Airpower Journal*, July 1995.
56 Grau. L-W., Thomas, T. (1996) "A Russian View of Future War: Theory and direction", *Journal of Slavic Military Studies*, issue 9.3 (September), pp. 501–18. Quote: "... from a military point, the view of Information Warfare against Russia or its armed forces will categorically not be considered a non-military phase of a conflict whether it will be causalities or not ... considering the possible catastrophic use of information warfare means by an enemy, whether on economic or state command and control systems, or on the combat potential of the armed forces ... Russia retains the right to use nuclear weapons first against the means and forces of information warfare, and then against the aggressor state itself."
57 CRS Report for Congress. "Cyberwarfare". Updated 19 June 2001.
58 Tsymbal, V.I. (1995).
59 Fitzgerald, M. (1996) "Russian Views on Information Warfare". *Hudson Institute,* Washington D.C. USA. December 1996.

In 2007, one proposed definition for information warfare offered by military theorists attached to the General Staff of the Armed Forces was the following:

> [The] main objectives will be to disorganise [disrupt] the functioning of the key enemy military, industrial and administrative facilities and systems, as well as to bring information-psychological pressure to bear on the adversary's military-political leadership, troops and population, something to be achieved primarily through the use of state-of-the-art information technologies and assets.[60]

Information warfare, in the Russian view, is conducted in *peacetime*, in the *prelude to war* and in *wartime* on three levels – strategic (the state level involving different ministries and agencies, as well as operations on two or more fronts), *operational* (a front, an army, a corps) and *tactical* (a combined unit, a subunit).[61] In the Russian Armed Forces, IW consists of *electronic warfare, psychological operations, reconnaissance* (intelligence), *deception* and *mathematical programming impact*.[62] It should be stressed that the current definition does not explicitly mention computer network operations (CNO) but the term "mathematical programming impact" probably involves offensive and defensive capacities for computer and network exploitation, attack and defence.

In peacetime, information warfare is related to information security for society and the government, including a wide range of aspects that have to do with protecting the state. Information warfare is conducted secretly by means of intelligence, politics and psychological operations. At international level it involves diplomatic and economic measures and methods.[63] At supranational level the objective of special information operations (SIO) is to shape public opinion (at home and internationally) and thwart alliances between possible adversaries.[64]

Maskirovka (methods for deception) is a constituent element in peacetime IW. It is an element of stratagems which "control" the enemy by creating

---

60 Dylevsky, I.N., Komov, S.A., Korotkov, S.V., Rodionov, S.N., Fedorov, A.V. (2007) "Russian Federation Military Policy in the Area of International Information Security: Regional aspect", *Moscow Military Thought*, 31 March 2007, referenced by Carr, J., 27 July 2009.
61 Limno, A.N., Krysanov, M.F. (2003) "Information Warfare and Camouflage, Concealment and Deception". *Military Thought*, vol. 12, No. 2.
62 Limno et al (2003).
63 Pirumov, V. (1996) "Nekotorye aspekty informatsionnoi voiny" (Certain aspects of Information Warfare). Conference presentation in Brussels May 1996, referenced in Thomas (1998a).
64 Donskov, Y., Nikitin, O.G. (2005) "Special Information Operations in Armed Conflicts". *Military Thought*, vol, 14, No. 3.

a false impression of the actual situation, the status of the forces opposing the enemy and about the concept, time and nature of their operations, forcing him to act in a predictable manner that will be unfavourable to himself.[65] The Russian Armed Forces' experiences of *maskirovka* are very good.

In wartime, information warfare refers to achieving information superiority (information dominance) over the enemy, to gain and maintain information advantages but also to protect a country's own information and information systems.[66] IW operations in wartime are more overt than in peacetime and can support traditional forms and methods of warfare, including information and intelligence activities.

Information warfare weapons consist of "... any technical, biological or social means or system that is used for purposeful production, processing, transmitting, presenting or blocking of data or processes that work with the data."[67]

They involve the physical destruction of military information systems, electronic countermeasures, specially programmed hardware and software (interpreted as malware, such as viruses, worms, trojans, back-door functions, logic bombs), and the distortion, deception and manipulation of information, including psychological operations.[68]

Alexandr Burutin, deputy chief of the General Staff, has made the following statement:

> Information weapons [...] do not require specialised manufacturing facilities and a complex infrastructure. A small group or even one expert can develop and carry out an act of destruction while not having to physically cross borders and expose human lives to risk.[69]

The statement may mean that even a small number of skilled and dedicated hackers can inflict great harm on an enemy's critical systems. Different kinds of viruses and malicious code are used for a negative impact on information systems and to induce detrimental behaviour. Other methods include illicit information collection and cyber espionage. Kukashkin and Yefimov have addressed concerns about hostile actions in the form of "algorithm bombs"

---

65 Fitzgerald, M. (1996).
66 Pirumov, V. (1996).
67 Rastorguyev, S.G. (1998) "Informatsionnoi Voiny" (Information warfare). Radio in Svyaz, referenced in Thomas, T. (2004) "Russian and Chinese Information Warfare: Theory and Practise". Foreign Military Studies Office, Fort Leavenworth. PowerPoint, June 1998.
68 Rastorguyev, S.G. (1998).
69 Presentation at Info-Forum, 10 February 2008, referenced by Jeffrey Carr in AppSec Asia Conference, 17 November 2009.

and "software bombs". This kind of malware can distort a section of an algorithm and limit the functionality, thus causing unreliable behaviour.[70]

The Russian arsenal also includes radio frequency weapons and "psychotropic weapons", i.e. mind control weapons, with the purpose of influencing neurological activities in the human brain. Whether this kind of weapon really exists or not is a subject of dispute among Western analysts.

Other arms include electromagnetic pulse weapons that knock out electronics and components at great distances – temporarily or permanently.[71]

The Chinese approach

Just as Russia, China lacks an official doctrine on information warfare and information operations, one that corresponds to that of the U.S. On the other hand, a number of influential persons in People's Liberation Army (PLA) and other organisations have described the phenomenon and developed and adapted theories in the field. These theories are assumed to form a common base and starting point for a specific Chinese capacity.

The Chinese view is to some extent based on interpretations of the U.S. IO doctrines when it comes to a network centric approach and information superiority. However, it has been adapted to the Chinese situation, environment and culture to become a specific strategy of its own. Just as for the Russians, the Gulf War in 1990–91 is a watershed and strong influence for China. The United States and its modern armed forces is the adversary that the nation has to relate to. Today, U.S. military power is superior to China's, at least when it comes to conventional forces and nuclear weapons. On the other hand, China's perspective is very long, their plans to modernise the armed forces stretch beyond 2050.

The basic elements in the Chinese view on information warfare are *asymmetric warfare* (inferior versus superior), *preemptive strike capability, first strike attack, high-tech local wars* and *popular war*. In some documents the term "unlimited warfare" has been mentioned as part of the Chinese strategy on information warfare. The term is disputed by several analysts.

---

70 Kukashkin, A.N., Yefimov, A.I. (1995) "The Security of the Infosphere of Strategic Defence Systems". *Military Thought*, No. 5.
71 Thomas, T. (2004) "Russian and Chinese Information Warfare: Theory and Practise". *Foreign Military Studies Office*. Fort Leavenworth. USA. PowerPoint, June 2004.

Other important parts include the *stratagems* of military strategist Sun Tzu,[72] described in *The Art of War* from 500 BC.[73] The 36 stratagems involve *information gathering* and *deception*. These can compensate for any shortcomings in physical resources when facing a stronger enemy. Sun Tzu says that warfare is the art of misleading, and winning the conflict is vital:

"In war you have to win. In an attack you have to take the initiative. If there is no advantage for the nation, don't go to action."

"If you can't win, don't go to war."

"Deceive the adversary by pretending to follow his will."

Sun Tzu's ideas have influenced warfare in China for more than 2000 years, and they form the basis of modern intelligence. In no way does Sun Tzu advocate war; instead he says it only brings sorrow to people. Despite this, it may be necessary in order to survive in a hostile environment. Sun Tzu's thesis is that if a conflict cannot be avoided, good preparations are decisive in order to be successful. This is done through espionage.[74] It is equally important to analyse and draw the right conclusions from the collected information and use it strategically, tactically and operationally. This creates conditions for victory without having to fight and kill. The greatest of victory is to win the conflict without even going to war:

It is better to subdue a country in a complete and undamaged condition than to destroy it completely. The damages that the conqueror causes, makes the victory less impressive.
[...]
Greatest is not to win one battle after another. It is better to win through enemy capitulation without fighting.

Sun Tzu's ideas have been elaborated and adjusted through the centuries. One important modern Chinese military theorist is Dr Weiguang Shen.[75] He is described as a prominent figure in the field. Other important persons that have influenced Chinese attitudes include General Wang[76], Lieutenant

---

72 Sun Tzu lived in the kingdom of Wu during the "period of the warring states", which lasted 403–221 B.C.
73 Tsai i Chih Chung: "Suntzi talar. Krigskonsten" (1996) *Alhambras Förlag AB*.
74 Sun Tzu divides espionage into five categories; local spies, insiders, drop-out spies, expenditure spies and secure spies.
75 In 1985, the 25-year-old army soldier published a research report mentioning the possibility for China to conduct a new kind of warfare based on computers. In the paper "A New Form of People's War", dated 25 June 1996, Shen concludes his ideas.
76 Wang, B. (1995) "The challenge of Information Warfare".

Colonel Wang Baocun, Li Yinnan[77], Zhenxing and Li Fei[78] and Major General Dai Qingmin.

One criticism that several Chinese experts direct at the American doctrines on information operations is that they are too technology-driven and do not ascribe the strategic dimension enough importance.[79] There is too much focus on the enemy's information and information systems, and too little consideration of softer psychological factors. The Chinese theories tend to stress the importance of understanding cognitive processes in the way the enemy perceives the situation, rather than technical aspects. The purpose is both to understand, and to create possibilities to influence the enemy's will to fight.

Moreover, the Chinese view is not limited in time and space to a specific conflict or crisis; information warfare is continuous over long periods of time.[80] It involves the whole spectrum of phases in peacetime, crises and military conflicts.[81]

According to Dr Weiguang Shen there are two kinds of wars. One is violent; it takes place on the battlefield. That kind is limited geographically and time-wise. The other is multi-layered, with different intensities.[82] Information warfare, especially, is such a continuous conflict, according to Shen. The theory is in line with the Chinese long term view in order to achieve its comprehensive strategic goals.

A fundamental factor in information warfare is the capacity to influence the enemy's *decision-making process*. According to Liang, the essence of information warfare is to disturb, preferably prevent, the enemy's ability to make decisions while protecting one's own. Similarly, in 1999 Shen wrote that one of the fundamental factors in this kind of warfare is to force the enemy to surrender without fighting. This is achieved through information

---

77 Li, Y. (1996) "New Subjects of study brought about by Information Warfare". *Jiefangjun bao*, 1996.
78 Wang. B, Li, F. (1995) "Information Warfare".
79 Thomas, T. The information is collected from a number of power point slides from a presentation made at the Swedish National Defence College in 2004 describing the Chinese and the Russian views on Information Warfare.
80 Barret, B.M. (2008) "Information Warfare: China's response to U.S. technological Advantages". *International Journal of Intelligence*, vol. 18, No. 4.
81 Wang, B. (1997) "A preliminary Analysis of IW". *China Military Science*, No. 4, 20 November 1997.
82 Barrett, B.M. (2008).

superiority. There is a clear link from Liang's and Shen's statements to theories on network centric warfare and Sun Tzu's stratagems.
American analyst B.M. Barrett has tried to interpret the different military theorists' disparate views and ideas, and sum them up in a common Chinese definition of information warfare. His interpretation is:

> Information warfare is the struggle with and about information resources that appears both physically and cognitive. Actions and activities happen through the whole political-military spectrum in peacetime, during time of crisis and in wartime. Both offensively and defensively methods are being used in order to influence the adversary's will and capability to act in a conflict.[83]

On 20 August 2000 Major General Dai Qingmin wrote in the journal *China Military Science* (*Zhongguo Junshi Kexue*) that information operations are a product of circumstances, goals and forms. Qingmin's interpretation is a:

> ... series of operations, where the information environment is the battle ground and military information and information systems are the operational targets. Electronic warfare and computer warfare are basically directed against the adversary's strength and knowledge.[84]

Two years later, in the same journal, he pointed out that information warfare consists of the basic elements operational security, deception, psychological operations, electronic warfare, computer and network operations and physical destruction. The focus for the PLA is to develop the capacity to integrate electronic warfare and computer network operations – Integrated Network Operations and Electronic Warfare (INEW).[85] Dai Qingmin adds that INEW is both an "over all concept, a method and a strategy for information warfare". It is about protecting one's own information and intelligence systems as well as destroying the opponent's. Dai says that in digital warfare an enemy should be seen as a system, not as individual parts such as platforms, individuals, weapons systems etc.[86]

Through electronic warfare, such as electronic corruption of signals, it is possible to interrupt the reception and transmission of information. By using computer and network operations the process to refine data to infor-

---

83 Barrett, B.M. (2008).
84 Major Dai's view on IW/IO is described in power point slides made by Thomas 2004.
85 INEW – Integrated Network Operations and Electronic Warfare. See Rogin, J. (2006) "China fielding cyber attacks units".
86 Power point slides made by Thomas, T. 2004.

mation and knowledge, and use it, is denied. That way control over the information arena has been reached.[87]

Chinese military analysts have implemented several of Sun Tzus's 36 stratagems and adapted them to conditions in cyberspace as information warfare stratagems. Major General Niu Li, Colonel Li Jiangzhou and Major General Xu Dehui in an article in *China Military Science* in August 2000 described a number of ways for the Chinese military command to apply ten stratagems to the information arena in order to achieve long-term strategic goals.[88] The purpose of these is to implant "faults" and corrupt information in the enemy's systems, to influence contents and processes as well as the thought patterns of influential people.

In line with Dr Shen's ideas, the interpretation is that at strategic level there is an aim to undermine the enemy's capacity to understand a situation correctly and counteract his ability to handle the situation by making the right decisions. The authors also write that another purpose of the stratagems is to divide the enemy. This can be done by directing a number of different multidimensional information threats against him.

Such a division leads to a dilution of resources for the party under attack, resources that have to be distributed between the threats, resulting in weaker links and gaps. This, in its turn, increases the aggressor's freedom of action to make new assaults. The aggressor wins in strength by combining different means of information warfare in time and space, concentrated on the enemy's weak points. Viruses and malicious code are used to blur the opponent's flow of information. Deception, false and corrupt information, false targets and concealment of intentions are fundamental and basic elements in the stratagems. Operations should be conducted in a methodological way and during a long period of time. The goal is to gain control over and survey the situation. It is important to make the right decisions; choosing the right time for an assault is a decisive factor.

According to Chinese analysts there is an imbalance between the competing parties in China's immediate surroundings when it comes to resources,

---

[87] Dai, according to Thomas, uses the term "network space" which is interpreted as the information domain.
[88] Niu, L., Li, J., Xu, D. (2000). "Planning and Application of Strategies of Information Operations in High tech Local War". *Zhongguo Junshi kexue* (China Military Science) No. 4, 20 August 2000. In Barret "Information Warfare: China's response to U.S. Technological Advantages". *International Journal of Intelligence*, vol. 18, No. 4.

access to advanced technology, manpower etc. The Chinese doctrine on asymmetric warfare is based on the idea that in a conflict the nation has to use its limitations as advantages in order to balance these injustices. The strategy is called *inferior versus superior*. Information warfare is an important tool and a specific capacity to perform preemptive strikes. In this case, asymmetric should be understood in relative terms, as it is not costly for an antagonist to conduct cyber attacks, while it is costly for the party that is attacked to defend itself.

One interpretation of the Chinese doctrine is that information warfare is asymmetric; information is in itself asymmetric in the sense it is used as the weapon of the "weak party". It is a relatively cheap weapon that can be used in surprise attacks in order to knock out the enemy's vital information and communications infrastructure.

A surprise attack against vulnerable targets may give an advantage, a momentum in time and space, which can be decisive for the outcome. The method advocated is distributed denial of service attacks, DDoS, against enemy systems, in the same manner as in Estonia in the spring of 2007 and in Georgia in the summer of 2008.[89] The U.S. weakness is its great dependence on digital command and communications systems that integrate different modern systems such as high-tech sensors and weapons systems. By attacking these, asymmetric effects can be achieved.

According to Triplett[90] the People's Liberation Army shows an interest in creating a preemptive strike capacity against vital enemy infrastructure. They invest heavily in capacity for offensive information operations, but also to be able to control information (and the information arena), both their own and the enemy's, as advocated by Major General Dai Qingmin and other Chinese experts.[91]

In 1993, after having studied the outcome of the Gulf War, former president Jiang Zemin declared that the focus for the PLA should be *local wars under modern, high-tech conditions*. It may be assumed to primarily de-

---

89 Nicoll, A. (2007) "China's cyber attacks. Casting a wider intelligence net". *IISS Strategic Comments*, vol. 13 issue 07, September 2007.
90 Triplett, W.C. "Potential Applications of PLA Information Warfare Capabilities to Critical Infrastructures".
91 Wen, T. (2002) "PLA Bent on Seizing Information Control", *Hong Kong Ching Pao*. 1 June 2002, in Barret "Information Warfare: China's response to U.S. Technological Advantages". *International Journal of Intelligence*, vol.18, No. 4.

scribe limited geographical conflicts within China's sphere of interest such as Taiwan and the South China Sea.

Warfare in the information arena or in space[92] is not restricted to a certain land area or arena in the same as with physical attacks; they can be carried locally, regionally and globally. There are no boundaries in cyberspace; an attack can be initiated tens of thousands miles from the target and have effects on both the enemy and third parties. One interpretation is that the Chinese information weapon is viewed as a "force enabler" in a potential conflict situation. The weapon will be used globally and across borders against adversaries, both to throw them off balance and to create momentum and a regional advantage.

Due to its Taiwan policy, the United States is seen as China's primary adversary. India, Russia, Japan and South Korea are also potential enemies. The United States, just as other Western countries, depends on working information technology in order to maintain its civilian and military structures. IT is a strength for the country, but in this case also a weakness. It is vulnerable to information operations, particularly in the initial phase of a conflict. In a statement in 2000, Robert T. March, chairman of President Bush's commission on cyber security, warned of the risks of an electronic Pearl Harbor.

Barret has described a scenario where China, in a limited local conflict with Taiwan, tries to prevent the U.S. from getting involved through different preemptive information attacks against vital infrastructure in the country.[93] There would be severe problems for the U.S. In the scenario, the Americans fail to redeploy troops to the conflict zone in due time. In this case it gives China an advantage and superiority versus its opponents. China gains control over both the physical arena in the conflict area and in cyberspace. The United States backs down, and China fulfils its policy against Taiwan.

Chinese analyst Wei Jincheng[94] claims it is possible to increase the effect of an operation by combining information warfare with the classic Mao-

---

92 The space arena can be viewed both as a physical domain to be used in Navigation Warfare (NavWar) and as part of the information domain.
93 Barret, B.M (2008).
94 Jincheng's ideas to conduct virtual mass attacks are described in Thomas, T. (2004) *Chinese Information Warfare Theory and Practice*. Foreign Military Studies Office (FMSO), Fort Leavenworth, Kansas, June 2004.

inspired Chinese doctrine based on a popular mass army, a reserve army to be used if and when necessary.

American intelligence analysts have interpreted this as China being prepared to apply the mass army idea in cyberspace, in case of a military conflict between the nations.[95] The plan is to provide hundreds of millions of people with laptops connected to the Internet in order to conduct massive cyber attacks against vital U.S. infrastructure. Centrally controlled botnets would probably make it easier to co-ordinate the attacks from such a mass army.

Whether the strategy can be applied or not is open for debate. Logistically, it seems very complicated to realise, especially when thinking of the enormous demands for training of people *en masse* for this purpose. Another question is what effect really could be achieved. The task for this kind of reserve units will probably be to defend and secure Chinese IT infrastructure rather than to conduct cyber attacks against opponents.

It is also important to ask oneself whether the Chinese leaders really would want to put "the information weapon" in the hands of a large part of the population – especially when thinking of the overwhelming surveillance apparatus of Chinese citizens and the way the government censors information on the Internet. This is a double-edged sword, and definitely dangerous for the regime. Realising a project of this kind would demand a lot from the general public when it comes to education and training in IT.

Difference and similarities

The three dominating superpowers have their own ideas and thoughts on how to interpret and use information warfare. As opposed to the United States, both Russia and China lack official doctrines on information warfare and information operations, at least as far as is known outside these countries. All three countries stress the important role information plays in today's conflicts. It is going to grow even more. The United States has influenced the thinking of the others, especially when it comes to ideas about information superiority, information dominance and command and control warfare. Information adds a new dimension to warfare and IW weapons can be used both offensively and defensively, to protect a country's information resources and systems.

---

95 Barrett, B.M. (2008).

Russia and China take a broader view than the United States on the essence of information warfare in the sense that in their approach it covers both peacetime and wartime situations, while the U.S. definition is more narrow and related to times of crisis or conflict.[96]

The Chinese concept originates from Sun Tzu's 36 stratagems, described in *The Art of War*. One of the key factors in the Chinese concept is deception. The IW perspective covers a long period of time and is not limited to a specific moment, period or conflict. Chinese experts criticise the U.S. doctrine for being much too technology-driven and for not considering the strategic dimension sufficiently. Moreover it is too focused on the opponent's information and information systems and does not consider the softer, psychological factors. In the Chinese conceptual framework, cognitive elements are added, such as the opponent's *will and capacity* to fight. It has a clear political dimension. According to Sun Tzu, "To win the war without the fight is the greatest victory".

In the Chinese approach IO is a component of IW, contrary to the U.S. view. For American experts *IO is a way to fight* while the Chinese think that *IW is the fight itself* and it is waged on many different levels and in many different dimensions during a long number of years.[97]

The Russian view is closer to the Chinese; it stresses the information-psychological impact and the idea that IW is conducted in both peacetime, in the prelude to a conflict and in wartime. It is also more or less constant; at strategic, operational and tactical level.

When it comes to deception, all three see the term as a vital part of IW/IO. The United States uses the term *military deception* (MILDEC) as a core capacity of IO in order to mislead an enemy's decision makers (in a conflict situation). Compared to the U.S. view, the Russian armed forces treat *maskirovka* as an independent kind of operational (combat) support to affect an enemy. It is conducted on a daily basis and at all levels.

The time perspective is probably longer for the Chinese, covering several decades or more, compared to that of Russia. The U.S. time perspective is shorter than both Russia's and China's and it is related to specific conditions and conflict situations. On the other hand, the Americans occasionally use the term *strategic communication* in order to exert influence at strategic and political levels over a longer period of time. It is not necessarily limited to

---

96 Thomas, T. (1998a).
97 Barret, B.M. (2008).

a specific conflict. Strategic communication is not an integrated component of information operations. Moreover, it is important to note that MILDEC and strategic communication are not necessarily parts of one another.

The U.S. joint doctrine stresses that information superiority or information dominance is a key factor in order to create a joint capacity for command and control between the service branches. Information superiority is an important advantage. Russian analysts claim that information superiority will be the most important component for victory in 21st century wars.[98] Bogdanov says that "… it will be impossible to attain strategic and operational objectives in future wars without achieving superiority over the enemy in the information sphere".[99] The Russian and Chinese views are similar to the U.S. approach in that aspect. Another similarity, according to Thomas, is the concept of protecting one's own information while affecting that of an enemy.[100]

Information control, as defined in the Doctrine on Information Security of the Russian Federation,[101] has a built-in psychological dimension that has to do with the stability of the state. It is intended for the country's population as well as what is seen as foreign campaigns to exert influence. The Russian and Chinese views show some similarities, although their practices when it comes to how to gain control over the media, the Internet and other communications channels (for instance) often differ. In Western democracies the ability to restrict or control the media component and the information flow is regulated by laws.

In the Russian and Chinese concepts of IW, cyber operations are directly connected to psychological activities. They are integrated. On the U.S. side, the line between these does not seem to be as important as it is for China and Russia. The reasons could be both historical and organisational.

One major difference between the United States and Russia, and probably China, is their view on cyber attacks (e.g. computer network operations) against their own systems and the risk of an escalation of conflicts on the Internet. Some Russian analysts compare IW to weapons of mass destruction. In their view, if Russia's critical information and communica-

---

98 Limno, Krysanov (2003) "Information Warfare".
99 Quote: "… it will be impossible to attain strategic and operational objectives in future wars without achieving superiority over the adversary in the information sphere", in Bogdanov, S.A. (2004) "Warfare of the Future". *Military Thought*, vol. 13, No. 1.
100 Thomas, T. (1998a).
101 *Doktrina Informatsionnoi Bezopasnosti Rossiiskoi Federatsii*.

tions systems are attacked digitally, they reserve the right to use nuclear weapons against the attackers. This way of looking at things is controversial, and it is probably not the official Russian view. The Russian statement should be interpreted as deterrence. That, at least, is the message Russia wants to send to its potential enemies.

This is not a coincidence, but an accepted behaviour.[102] Russia's main deterrence is to show a potential enemy the cost of launching a cyber attack against the nation. Regardless of whether the threat is backed up or not, this shows the importance ascribed to information warfare. This applies to all the super powers, not just Russia.

---

102 Bogdanov (2004) "Warfare of the Future". *Military Thought*, vol. 13, No. 1.

Cyber terrorism. Electronic Jihad

The logic of terrorism is based on fear. It plays on the fear of unpredictable attacks; attacks that can take place anytime, hit sensitive targets and happen to all of us, without discrimination. Terrorism is about shocking people, making them feel powerless. The aim is for the common man to start questioning the foundations of society and the state's capacity to guarantee its citizens' safety.

For terrorists it is important that the operations are large-scale, with big effects, such as the attacks on September 11, 2001. They also have to be lethal and result in extensive media coverage. The latter is decisive – a condition both to make the political goals of the terrorists known and to demonstrate the strength of their organisations. Through advanced and co-ordinated attacks, a terrorist organisation can strike at a stronger enemy and preferably humiliate him. A successful operation can also be a fertile ground for recruitment of individuals and groups who are willing to sacrifice themselves in the future.

There are some similarities between terrorism and guerrilla warfare.[103] Guerrilla wars are by definition asymmetric. One party has more resources than the other, a technological advantage etc. This forces the weaker party to use unconventional tactics and vary his behaviour depending on the situation and conditions. It should be noted that asymmetric warfare is not necessarily the same thing as terrorism. However, terrorism can be used as a tactic by the weaker party in a conflict.

Terrorism in cyberspace

Cyber terrorism is a generic term for various activities in cyberspace; it involves a number of different organisations, groups and individuals. One

---

103 Giacomello, C. (2004) "Bangs for the Buck: A Cost-Benefit Analysis of Cyberterrorism", *Studies in Conflict & Terrorism*. University of Bologna, Italy. Taylor & Francis Group.

of the most well-known researchers on cyber threats, the American professor in information security Dorothy Dennings, defines the terms as illegal, socially or politically determined assaults or threats of assaults against computers, networks and stored information. For an attack to be regarded as cyber terrorism the intended effect of it has to be serious human and economic casualties, intense fear and anxiety – terror – among the citizens. If a cyber attack against vital information infrastructure shall be viewed as an act of terrorism or not depends on the intent.[104]

Players in the cyber arena may have the same political, ideological and religious objectives as "traditional" terrorist organisations, but they behave differently and work with different means. Co-ordination may take place within and between organisations. Cyber terrorists are not suicide bombers, nor is attention in the media an end in itself; on the opposite they usually try to hide what they are doing on the Internet as far possible.

In some cases, it is possible to distinguish between "pure" cyber terrorists and terrorists who use the Internet for co-ordination and attacks. However, the line between the two is getting increasingly blurred. The "pure" cyber terrorist does not manifest himself, instead he hides his intentions, at least until a full-scale infological attack has been performed. For the perpetrator, cyber terrorism is relatively safe, profitable and hard to detect.

Due to the fact that more and more systems are interconnected and become dependent on computer networks, new vulnerabilities appear that can be exploited by ill-intentioned individuals and groups. Some people say that the information revolution has inflicted a very risky electronic Achilles' heel on society. A popular expression to describe this new threat is the risk of an "electronic Pearl Harbor". Probable targets include command and control systems in critical infrastructure.

Information about good targets for attacks can be found on the Internet. This kind of data decides what sort of methods and cyber weapons are optimal and where there are vulnerabilities in the computer network. Terrorists can use the Internet as a way of both co-ordinating and communicating between different parts of the organisation.

The term cyber terrorism has media potential; it is titillating and appeals to a wide variety of feelings. The Israel-based security policy expert Gabriel Weimann says there are a number of reasons why fear of cyber terrorism is

---

104 Dennings, D. (2000) *Cyberterrorism. Testimony before the Special Oversight Panel on Terrorism Committee on Armed Services*. US House of Representatives, 23 May 2000.

increasing.[105] Apart from the media, this feeling has been kindled by political and economic forces. There is also a general ignorance among the general public about the actual possibilities and weaknesses of the technology; all this affects the general feeling of vulnerability and (in)security.

In academic discussions on cyber terrorism there are often two disparate views:

1. The threats are real and considerable, since vulnerabilities multiply when a number of different systems are increasingly integrated, controlled and run via computers. This is primarily the view of governments and security firms.
2. There is no threat at all or only very limited ones. There are no, or only very few, known cases of cyber terrorism. This view is primarily represented by scientists and analysts at universities and research institutes. They claim the concern is exaggerated; security systems generally work well thanks to the efforts made before the turn of the millennium. Additionally, it is in the interest of some security firms to add to the unease, since there is money to be made from it. Another argument is that the cost of cyber terrorism is greater than traditional terrorism. Cyber terrorism is not economically rational.[106] It's cheaper to hang rucksacks with explosives on a couple of suicide bombers, the media result is also greater.

It has also been pointed out that the Internet is used as a tool for coordination by terrorists who are planning operations. It would thus be unwise – counterproductive – to attack a system that benefits them.[107]

The validity of both parties' arguments can be discussed. The truth is somewhere in between and varies depending on the time and the place. Not every incident is made public. In fact, there are more or less serious cyber attacks every day in the form of hacking, distribution of viruses, thefts of passwords etc. A number of companies and governments all over the world have been subjected to penetration attempts. In some cases relatively harmless attempts, in other cases much more serious and costly for the victim.

A number of incidents are known that were probably staged by cyber terrorists, e.g. in the early 2000s when different kinds of computer viruses

---

105 Weimann, G. (2004) "WWW.Terror.Net: How terrorism uses the Internet". *United States Institute for Peace*, special report number 116.
106 Giacomello, C. (2004).
107 Christiansson, H., Fischer, G. (2003) "Terrorismens tid". *SNS Förlag*.

– such as *Nimda, Code Red, Love Bug* and later *Melissa* – were used. In many cases it is still not known who were behind the operations. The same is true of who the real targets were – and why the attacks were launched. The Internet makes anonymity possible, provided the antagonist is skilled and has sufficient knowledge to cover up his digital tracks. The consequences of these attacks have so far been limited and they do not follow the usual logic of terrorism: spectacular effects and high rates of casualties.

So far there has not been a single lethal cyber attack, at least not publicly. On the other hand, it is known that malicious code has caused serious accidents. For instance, in the 2008 crash at Madrid-Bajaras airport, one of Spanair's central computers was infected by trojans. If an aircraft sent distress signals with at least three technical problems, the central computer was supposed to raise the alarm. The airplane that crashed did send this kind of distress signal to the computer, but due to the trojan the alarm was not raised.[108] This had disastrous consequences; the accident caused 154 casualties.

It is not known who was behind the trojan, or what purpose it had. There is no proof that the crash was a terrorist attack. It may have been the result of a number of unfortunate circumstances that converged in the airline's failure to secure its central computers from external infection of malicious code. Irrespective of the cause, the crash shows the risks and consequences of cyber antagonism. This kind of event can cause unease among people that vital systems are not always sufficiently protected and reliable. This may in its turn influence the general public's trust in the authorities' capacity and preparedness to deal with risks, vulnerabilities and crises; something that may have political implications in the long run.

Cyber terrorism is not limited to organisations and individuals; states are sometimes also involved. One example is North Korea. Many nations in the international community define North Korea as a terrorist state; primarily it acts against its own citizens, but also against its main enemy, South Korea. In the spring of 2009 it was revealed that North Korean cyber forces had hacked into the South Korean National Institute of Environmental Research. They allegedly stole more than 2,000 secret documents, including the names of 700 companies that handle toxic chemicals.[109] Information from the intrusion could be used for terrorist attacks.

---

108 "Trojan bakom flygkrasch". TT, 21 August 2010.
109 "ROK Monthly Claims DPRK hacks, Steals Military secrets in March 2009". *Chosun Ilbo Online*, 19 October 2009.

Terrorists and other players may be developing methods and strategies to conduct large-scale digital attacks with deadly intent.[110] Examples of possible targets include financial systems, civilian air traffic, health care and energy systems, such as nuclear power plants.[111] This should be seen in the perspective that the costs of carrying out these kinds of operations keep decreasing.

A growing problem is intrusion attempts against databases in order to gather sensitive information on different kinds of objects, their data structures and information security levels. One may assume that groups like cyber terrorists have an interest in surveying vital infrastructure and physical targets that can be attacked. Another cause for concern is the risk of someone deliberately planting distorted information in sensitive databases, such as command and control systems.

Just as traditional terrorist attacks, cyber operations are carefully planned, with a clear idea of aims and effects. Often only a few people need to be involved in such attacks; the results may still be devastating.

Cyber terrorists mainly focus on civilian – not military – targets, since they are much more vulnerable. Looking at the consequences for society, the effect is much greater. Additionally, there will probably be more media coverage of a civilian target.

A trend since the middle of the first decade this century is that cyber terrorists have changed targets and modus operandi. The players have moved from generally quite simple methods to a very rational, sophisticated and purposeful behaviour. One example of the professionalization of cyber terrorism is al-Qaeda and their use of information technology in order to reach their ideological and political goals.

Jihad and al-Qaeda in cyberspace

Al-Qaeda ("the Base") is a terrorist network that was founded in 1988 by Saudi multi-millionaire Osama bin Laden. It is a transnational movement that stretches from Algeria to the Philippines, with sympathisers in the West. The organisation is responsible for a number of spectacular terrorist

---

110 Collin, B. (1997) "The future of cyber terrorism: Where the Physical and Virtual Worlds Converge", 11th Annual International Symposium on Criminal Justice Issues.
111 Heickerö, R., Hyberg, P., Olsson, G., Renhorn, I., Jonason, T., Eklöf, F. (2004) "Telekrig i en breddad hotbild". Underlagsrapport, FOI, December 2004.

attacks. September 11, 2001 and the 2004 Madrid bombings are the most infamous. Thanks to intelligence collection and incessant attempts by security services all over the world to dissolve the network, it has lost a great deal of its former capacity and been forced to transform and adjust its activities. The assassination of Osama bin Laden on May 2, 2011 was another serious blow to the organisation.

Today, al-Qaeda can be described as a franchise in the terrorism business. The core is assumed to consist of a couple of hundred people, scattered over various parts of the world. Except for these, there are a number of loose cells that act autonomously without direct orders from the inner circle and top leaders of the network. His sympathisers saw bin Laden as a charismatic leader and front, more of an inspiration and a financier than an organiser or administrator for the "cause".[112]

Jihad is what constitutes the Islamic political violence that al-Qaeda is a part of. Islam distinguishes between *Jihad Akbar*, "the greater holy war", and *Jihad Ashgar*, "the minor holy war".[113] The greater Jihad refers to the individual's struggle to become a good Muslim, characterised by self-preservation and self-control. The minor Jihad has political and military features and is regulated by a number of ethical decrees. It is about protecting Islam and its holy places. One decree says that Muslims can only enrol in (minor) Jihad when attacked. Modern Jihad is directed against the United States, Israel and their allies as well as Muslim states that support them and oppose the establishment of a "pure Islamic nation". The goal for al-Qaeda is to establish an Islamic caliphate stretching across the Arab world to Southeast Asia and covering all areas where Muslims live.[114]

During the years that have passed since September 11, the intentions and methods of al-Qaeda and other terrorist organisations have changed. The Internet has become an increasingly important channel for communication and recruitment, but also for information on suitable targets to attack.[115] The terms that have been used to describe this transformation include electronic, digital and cyber Jihad.

---

112 United Nations, Security Council Committee established pursuant to resolution 1267 81999, 20 September 2002.
113 Napoleoni, L. (2004) "Oheligt krig. Den moderna terrorismens ekonomiska rötter". Andersson Pocket AB. Stockholm.
114 Wright, L. (2007) "Al-Qaida och vägen till 11 september". Albert Bonniers Förlag.
115 Atwan, A.B. (2006).

Through the Internet it is possible to acquire means and methods to e.g. produce bombs for "traditional terrorism" as well as to develop the capacity for network operations. Since the Internet is not regulated and it is possible to spread information at a low cost, even small groups of players can gain a lot of attention for its views. It is easy to conduct psychological warfare and information campaigns. The Internet is used to reach ideological and political goals.

It is important to stress that the Internet is not a target in itself for radical organisations such as al-Qaeda. However, the infrastructure can be used by players with skills and resources to attack control and monitoring functions in systems vital to society.

The Internet plays a significant role for terrorist organisations such as al-Qaeda. This can be seen in the following quotes, drawn from a website called Jihad online, Islamic Terrorists and the Internet, which has now been closed:

> Thanks to the progress in modern technology, it is simple to spread information, news, articles and other information through the Internet. We express with the greatest conviction that young Muslims with good Internet knowledge should spread information and articles of Jihad. We continue our struggle with the help of Allah.[116]

In a raid of the Islamic organisation al-Muhajiroun and its leader Omar Bakri Muhammad in London in 2005, police found e-mail correspondence between Bakri and other known al-Qaeda sympathisers. One document contained the following text:

> Within the foreseeable future you will see attacks directed towards the Exchange. I would not be surprised if tomorrow I would hear about a major economic collapse because someone attacked important technical systems in big companies.

Omar Bakri Muhammad claims there are tens of thousands bin Laden supporters who are studying computer science in order to work for the holy cause.[117] The statement has been confirmed by Hamid Mir, terrorism expert at and editor of the Pakistani daily newspaper *Ausaf*. According to Hamid, bin Laden said in an interview that …

> … hundreds of young men have turned to him and pledged to die for the cause, and hundreds of Muslim scientists are with him and they'll use their knowledge in chemistry, biology and other fields, from computers to electronics, against the infidels.

---

116   The quote was found on the now closed websites www.jehad.net and www.jihadunspun.net.
117   Atwan, A.B. (2006).

Bin Laden allegedly also told his sympathisers that it is very "important to hit the American economy, on which they base their military strength". His statements could be interpreted both as psychological propaganda and as an expression of the organisation's intent to carry out operations in new ways with new means, methods and targets.

Targets of specific interest for computer and network operations include command and control functions for systems critical to society; SCADA systems.[118] Conducting successful operations requires a high level of skills and great resources, either from the organisation or a state player who wants to participate in an electronic Jihad. It is not clear whether al-Qaeda presently has these kinds of assets and contacts.

According to unconfirmed information, in 2006 U.S. intelligence found a lot of information in laptops in Afghanistan with photos, blueprints and documents describing a number of sensitive targets for attacks in the United States and Europe. Some of the information in the documents was based on Google Earth and other open sources on the Internet. It is obvious that Google Earth and similar websites are used in the planning of terrorist attacks.

A lot of information about vulnerabilities in networks and computers can also been procured via hacker forums. In Pakistan and other countries in the Middle East, and in the West, there are people with a lot of knowledge of how to penetrate computers and networks. There are a number of hacker groups that say they support the Islamic cause, including the Taliban and, presumably, al-Qaeda. Palestinian, Turkish and Moroccan groups, to mention just a few, are very active and have good qualifications in the field.

The cost for information on how to infect computers through spyware, which can be used for things like distributed denial of service attacks, is somewhere between 1,000 and 5,000 U.S. dollars.[119] On a prominent, but now closed, website associated with al-Qaeda there was information on how to conduct DDoS attacks.[120] There are a number of other websites with links to the Islamic movement that teach hacking and how to handle viruses and trojans.

---

118   SCADA – Supervisory Control and Data Acquisition. In several reports for instance the CSR Report for Congress. "Botnets, Cyber crime and Cyber terrorism: Vulnerabilities and Policy Issues for Congress". Updated 29 January 29, 2008, and in Dorothy Dennings works on cyber threats, anxiety are forwarded regarding al Qaeda's and other terrorist groupings interest to attack vital infrastructures.
119   Francis, B. (2005) "Hackers sell their information anonymously through secretive websites. Know thy hacker". *Infoworld*, 28 January 2005.
120   Greenemeir, L. (2007) "Electronic Jihad App Offers Cyber terrorism For the Masses". *Information Week*, 2 July 2007.

Due to its simplicity, structure and global penetration, the Internet has become one of the most important channels for terrorist organisations like al-Qaeda. The technology makes "networking" possible between the organisation's different and loosely tied parts. It is used as a primary tool to collect funding and find new sympathisers and recruits to Jihad. The global network is also used for psychological operations, disinformation and deception campaigns, as well as for blackmail of opponents. The Internet is an effective means of "self-radicalisation" of individuals and groups all over the world.

Illustration 1. Al-Qaeda's first website www.alneda.com, launched in 1997.

According to the Washington-based Institute of Peace, the terrorist network controlled more than 5,000 websites in 2006. At least 50 of these were in turn connected to a number of other forums.[121] Islamism expert Abdel Bari Attwan says there are more than 4,500 sites that help al-Qaeda develop into

---

121 Weinmann, G (2006) "Terror on the Internet: the new arena challenges". U.S. Institute of Peace, Washington D.C. 2006.

a truly global ideological movement.[122] Every year hundreds of new sites are added.[123]

It is hard to estimate the exact number of websites directly or indirectly associated to the terrorist network. They differ in structure and design, what kind of message that is presented and to what audience. Some of them are more associated with convinced "jihadists", others are linked to chat rooms and Internet forums. The interest al-Qaeda shows in the Internet is in no way new. Their first website was www.alneda.com, launched in 1997 on servers in Singapore, Malaysia and Texas, the United States. The Arabic word *Alneda* can be roughly translated as *the message*.

A couple of months after it was launched, the website was discovered by American intelligence and forced to shut down. Shortly after this it was re-launched, but under a new name and on different servers. Subsequently, Alneda changed both URLs and names every five days to avoid discovery. This cat-and-mouse game continued for more than six years, before the website finally disappeared from the Internet – for good it was assumed. But in April 2003 a website with links to al-Qaeda sympathisers was discovered; it carried the message: "Faroq continues to carry the banner of alneda.com."[124] Another early website was *Maalemjihad* – the holy war's milestones. It was launched on a primary server in China around 2000, with a mirror server in Pakistan.[125] Another important site, according to American security expert Michael Knapp, was *al Ansar*.[126]

Constantly moving material around on the Internet and launching it under new names on web hotels at different geographical locations is a strategy that has worked well. Between January 2002 and April 2003 al Ansar changed its websites approximately every other week. On an English forum, discovered in November 2007, there were manuals with instructions on how to produce "dirty bombs", both nuclear and biological versions. One of the manuals contained more than 80 pages of descriptions and quotes that were said to originate from Osama bin Laden. One of the quotes was: "The manual should be seen as a gift from the commander to the jihadist fighters." When the website was discovered it had had more than 57,000 unique visitors.

---

122 Atwan, A.B. (2006).
123 Weimann, G. (2006).
124 www.faroq.org.
125 Atwan, A.B. (2006).
126 Abu-'Ubayd al Qurashi, "the War of the Ether" on the web site al Ansar 20 November 2002. The quote is described in PowerPoint slides made by Michael Knapp, "The War of the Ether: Al-Qaeda's PSYOPS campaign Against the Western & Muslim Worlds".

There are a number of websites on the Internet that are more or less closely connected to al-Qaeda. They provide advice on how to design bomb belts used in suicide attacks. It is also possible to learn about tactical behaviour and how to use RPGs[127] against different targets. This kind of website can be seen as an information library and part of what is called the *Encyclopedia of Jihad*. The Taliban's official unit for media distribution shows detailed descriptions on how to produce and pack explosives. There are also interviews with the leaders of al-Qaeda talking about Jihad. The contents gradually change; new material is constantly added, often related to current events.

Just as new websites are continuously being added on the Internet, others are shut down by operators after pressure from law enforcement agencies. Other reasons why they disappear include hostile hacking activities by people and organisations opposed to the Islamist struggle and electronic Jihad.

Up until 2007 *Internet Haganah,* a Zionist organisation based in the United States, identified and confirmed at least 700 websites with terrorist material to Internet service providers.[128] Shortly after this, the websites were shut down. The organisation has probably also hacked these kinds of websites.

There are also a number of fake websites with messages that seem to come from al-Qaeda and the Islamist movement. These are used as "honeypots" in order to monitor possible visitors and catch their IP addresses. Through the IP address it is possible to trace and identify from what server the visitor initiated the session. Alneda's website has allegedly been repeatedly hoaxed. Who is/are behind these operations is not known. Generally speaking, with the right will and motivation it is quite easy to produce fake websites.

One of the more interesting websites when it comes to form and content is al Muhajiroun.[129] The site has the same name as one of the more prominent British Islamist movements, led by Sheikh Omar Bakri Muhammad. In the beginning of 21st century it became something of a role model in the art of propaganda and recruitment on the Internet. It is also a source of self-radicalisation. There is information on the website that emphasises the obligation of the faithful to kill non-Muslims. The website has now been removed from the Internet.

---

127 RPG – Rocket Propelled Grenade.
128 Http://internet-haganah.com/haganah/index.html.
129 Http://www.danielpipes.org/rr/1659.php.

Illustration 2. Al Muhajiroun, an Islamist recruitment website.

According to Islamist expert Abdel Bari Attwan, al Muhajiroun and similar sites often resemble one another.[130] They are divided into sections. The most important and prioritised is the religious one, called *fatwas*. It lists legitimate targets for attack, according to the views of the Islamists behind the site. These include people and organisations as well as physical objects such as the banking system etc. There are continuous references to Jihad. Visitors who want to learn more about the struggle can click on links on the website to get into contact with religious authorities, the *sheikhs*.

Fresh recruits are very important. In the Jihad section there is often material describing what anyone interested can do in order to become a holy warrior in the service of Islam. There are e.g. practical advice on how to get to different war zones in Iraq, Pakistan, Somalia and Syria; names and locations of mosques in different countries that are part of Jihad. There are photos and films describing successful attacks on the infidels by Jihadi

---

130   Atwan, A.B. (2006).

warriors, along with stories on heroes and martyrs who fell in battle. Their testimony to their comrades can be read and meditated upon. And of course there are donation links for those who want to give money to Islamist causes.

These kinds of websites in turn provide links to other websites and various Internet forums where there is more information. An interesting detail is that many of the websites have an IT section, where those interested are asked to share their knowledge of and information on how to develop new methods for electronic Jihad. From a modern marketing perspective this is a new way of making the "customer" get involved in the movement.

Along with mosques and universities, websites are presently probably the most important channel for propaganda in Western countries. Sheik Omar Bakri Muhammed has said that every year an average of 18,000 British-born Muslims are recruited for military training in countries where Islamist groups are active.[131] The number is probably highly exaggerated, but there are definitely recruitments. There are boot camps in Mali, Iraq, Pakistan, Somalia, Yemen, Indonesia and the Philippines. A number of Europeans have e.g. been recruited by Islamist movements in Pakistan and Somalia. Several of these Jihadists have been killed in fighting in Africa and the Middle East.

One website that was described by CNN as a propaganda centre for al-Qaeda is the English-language *Inspire*.[132] It is operated by American citizen Samir Khan, who is of Saudi extraction. Taimour Abdulwahab, who died in the failed terrorist attack in Stockholm at Christmas 2010, is reported to have been in contact with Khan via Facebook.

A well-known website for propaganda and recruitment is the British *Islamic Wakening*.[133] The website is linked to an organisation called *Azzam Publications*[134] which publishes Islamist material of different kinds. The interesting thing about Islamic Wakening is that one of the founders, Abdul al-Qari, is a man alleged to be a former Nazi, who used to go by the English name of David Myatt before he converted to Islam. The information has not been confirmed, but if it is true it shows that people with radical beliefs are not necessarily bound to their choice ideology; the importance thing is the struggle against somebody or something. According to information on the website they help organise resistance cells for Jihad.

---

131 MacCrory, D. (2000) "UK Muslims Volunteers for Kashmir War". *The Times*, 28 December 2000.
132 Cruickshank, P. (2010) "U.S. citizen believed to believed to be writing for al Qaeda website, source says". *CNN News US*, 18 July 2010.
133 Http://forums.islamicawakening.com.
134 Http://www.globalterroralert.com/azzam-mokhtar.pdf.

There are also websites associated with different hacker groups that are said to support al-Qaeda. One of these is OBL Crew – Osama bin Laden Crew. Islamist Hackers, a.k.a. Afghan Hackers, are believed to be part of the same group as the one that leads OBL Crew.

Recruitment is by no way restricted to men; there are sites specifically aimed at women. One is named after the woman poet al Khansa, who in the 7th century, in the early stages of Islam, wrote intrinsic complimentary texts to Muslim martyrs killed in the struggle against the infidels. The site encourages women to support their husbands in Jihad but also to introduce and prepare their children to continue the legacy. This is a duty. Naturally there is information on how to become a warrior in the service of Islam by constructing suicide bomb belts.

One effective way of spreading propaganda is to participate in chat groups and on free Internet forums. These forums are open in the sense that it is possible to express one's opinions; the subjects can be more or less anything, legal and illegal – hacking, Internet security, drug use, smuggling narcotics, music or fashion. Quite a few of these forums are superb tools for propaganda and recruitment of sympathisers, even co-ordinating terrorist activities for Jihadists.

Since the Islamist movement is global and resources vary, bandwidth is not supposed to be an obstacle for anyone interested. People from areas with poor IT infrastructure should still be able to download and access information. Accessibility is thus an important part of electronic Jihad. All members assume cover names and aliases, which they use when writing pieces or answering questions. Contacts between users are made via the forums.

Closed user groups are formed in order to avoid discovery by law enforcement agencies. The only way to be invited is through recommendations. When this obstacle has been passed, peer-to-peer connections are made with the other participants in the chat group. Passwords are exchanged between the parties and communication is enciphered.

Visits on these kinds of forums and chat sites are not free of risks; participation can attract attention from the police and prosecutors. An example of this is Scottish student Mohammed Atif Siddique, who was sentenced to a total of eight years in prison by a British court for having provided and distributed terrorist material on the Internet and in different forums.[135]

---

135 Lettice, J. (2007) "Jailed terror student "hid" files in the wrong Windows folder and provided terror instruction via web links". *The Register*, 23 October 2007. The information is based on an article published in the Scotsman.

Through computer forensics, i.e. criminal investigations of his seized computers and servers, British police could show that Siddique had hidden material that encouraged terrorism in a Windows folder. In a search of his apartment, and of properties and shops nearby, police also found more than 30 computers, 500 CDs and DVDs with compromising material along with 25 cell phones and 19 SIM cards. Almost 700 documents could be traced from the computers, with more than a thousand appeals for Jihad.

Naturally, this causes concern within Islamist ranks. There are continuous online discussions as to whether foreign security services have infiltrated the forums and found compromising material and tips on people and groups preparing for terrorism. Many participants are worried that the security services actively participate in the chat forums, asking questions and providing information in order to get a picture of the situation. According to Attwan, the Saudi intelligence services, for example, have allegedly shown an interest in activities on different forums.[136]

One of the most serious threats, when it comes to hostile activities, is from insiders. That is, politically and ideologically driven people with computer skills, who have been recruited and placed inside an organisation in order to conduct detrimental activities. For groups such as al-Qaeda the value of these people is great, so are the difficulties for police and security services to discover the moles. Below there is an example of a person who may have been acting as an insider, though this has not been confirmed.

On a blog discussing different aspects of terrorism, one case was brought up describing the security officer responsible for computers and networks at the University of Washington, School of Nursing in Seattle, Majid al-Massari. He was arrested in August 2004 accused of preparing for terrorism.[137] What he was actually charged with is not known to this author, but it probably had something to do with his job at the university. Majid al-Massari is allegedly the son of Dr Mohammed al-Massari, head of "the Committee for Defense of Legitimate Rights" (CDLR). Mohammed al-Massari is also the spokesman of al-Qaeda in London and known for his extremist views. Among other things, he has hosted websites with videos showing decapitations of infidels. Majid's stepmother, according to the blog, has participated in spreading the message of al-Qaeda in the United States.

---

136 Attwan, A.B. (2006).
137 Cochran, A. (2006) "Muslim Hackers Assaulting Websites since Cartoon Controversy Began". *Counterterrorism Blog*.

An anti-Islamist website describes the concerns of the British counter-espionage agency MI5 that important research laboratories handling biological, chemical and medical products have been infiltrated by al-Qaeda sympathisers.

A 2005 intelligence report by the U.S. Department of Homeland Security describes how the criminal group Jamaat ul-Fuqura, based in Pakistan, co-operates closely with an organisation called Muslims of America. Through its contacts with Jamaat ul-Fuqura it is believed that Muslims of America gained connections to al-Qaeda. They are also suspected of having provided members of Jamaat ul-Fuqura with information on hardware and computer networks.

There are also suspicions that associates of bin Laden used sophisticated insider information to speculate on the stock exchange prior to September 11, 2001. In the aftermath it was concluded that prior to the attacks there had been an unusually high level of trading in shares in the air transport, energy and insurance sectors. In a similar way, British intelligence, in connection with the subway and bus bombings in London in the summer of 2005, identified an increased activity in international computer communications between the suspects and IP addresses in various Arab countries.

The line between terrorism and "ordinary" criminality is vague. According to a 2008 U.S. Congress report, cyber criminals have entered alliances with drugs smugglers in Afghanistan and the Middle East.[138] The criminals supply the terrorists with stolen credit cards and passwords, in return they get access to narcotics.

Criminals have good skills in advanced information technology. Drug dealers use encryption intensely, in order to protect their communications over the Internet. Through their resources, they can hire computer specialists who help them with methods to conceal information, for instance in digital photos sent in seemingly harmless e-mails. The well-developed infrastructure in the West is a good facility both for co-ordination of transactions and logistics, but also for laundering money from criminal activities.

A report by the DEA, quoted in the 2008 Congress report, says that 14 of 23 identified terrorist organisations were involved in drug trafficking in 2004. The Islamist organisations were especially active. The annual turnover from narcotics in Afghanistan is estimated to be more 30 billion U.S.

---

138   CSR Report for Congress. "Botnets, Cyber crime and Cyber terrorism: Vulnerabilities and Policy Issues for Congress". Updated 29 January 2008.

dollars.[139] Drug money from the area, together with contributions from sympathisers, have become an important source of revenue for the Taliban and al-Qaeda. This should be seen in the light of the total annual turnover from narcotics, which is assessed to be 400 billion U.S. dollars.[140]

Provided that the information is correct, this would imply that al-Qaeda, the Taliban and various criminal groups have nearly 10 percent of the world's total narcotics turnover. Good financial assets make it possible for terrorist organisations and other criminals to hire computer specialists who produce ciphers that are very hard to crack. Discovering these activities thus becomes all the more difficult for law enforcement agencies.

A lot of the specialists are said to come from parts of the former Soviet Union and the Indian subcontinent.[141] Many of them probably do not know that they work for terrorists and/or criminal organisations; they have in all likelihood been given false information by their employers. In other cases they may sympathise with the political-ideological and religious intentions and goals.

Al-Qaeda is well-versed in ways of acting anonymously on the Internet in order to conceal their operations. In the planning for September 11, they used the Internet and e-mails to a high degree; for instance, communications between the terrorist cells inside and outside the United States were based on Internet telephone applications.[142] Khalid Sheikh Mohammed, one of the brains behind the attack on the World Trade Center, communicated with at least two of the hijackers via a special Internet chat software. Ramzi Yousef, sentenced to life in prison for the first attack on WTC in 2003, enciphered data in order to make it more difficult to discover his activities. Computers found in Afghanistan show that al-Qaeda collects information on targets via the Internet and sends enciphered messages about this.[143]

Cyber terrorism can be understood as a development of "traditional" terrorism with other means, but similar political and ideological agendas. Terrorists on the Internet take advantage of modern societies' increasing dependence on computer networks and mobile communications.

---

139 Handley, L (2007) "Intel Brief: Afghan Poppies Fear Not". *Center for Security Studies and Conflict Research, Swiss Federal Institute of Technology*, 2007.
140 Napoleoni, L. (2004).
141 CSR Report for Congress. (2008).
142 CSR Report for Congress. (2008).
143 Thomas, T (2003) "Al Qaeda & the Internet: the Dangers of Cyber planning".

What kind of conclusions can be drawn from the development? One is that Internet has become an important tool for terrorist organisations such as al-Qaeda to conduct their attacks. They act both in the physical world, with suicide attacks and various kinds of criminal activities, but also in the virtual world, with a large amount of websites. Since September 11, 2001 al-Qaeda's operational capacity on the Internet has increased. It involves information collection on sensitive targets to attack, co-ordination and recruitment. Cyber terrorism and electronic Jihad are a reality today.

One identified risk is that the terrorist network will not limit its activities on the Internet to communications and messages, but increase its ambitions to qualified operations against websites, servers and networks. This could be combined with physical attacks. Cyber assaults should be viewed as a force multiplier to physical attacks. It is uncertain whether the movement presently has that capacity.

In connection with the assassination of Osama bin Laden in Abbotabad in the spring of 2011, American intelligence services found a number of computers with sensitive material. It turns out that the terrorist network planned to attack the American railway system. How far they got in the planning process is not known.

Since al-Qaeda to a high degree co-operates with the Taliban in Afghanistan and Pakistan, as well as with al-Shabaab in Somalia, there is a risk that knowledge about cyber operations will spread to these organisations too. This could include how to carry out DDoS attacks, encipher communications or stay concealed on the Internet in other ways.

Industrial espionage and theft of information

The difference between espionage in the traditional sense and industrial espionage is marginal. Both are conducted in similar ways and with the same methods, although targets may differ. In principle, this is the world of organised theft of information. Most countries and companies look for information on their opponents and competitors. Even friendly nations and companies can be of interest. Using open source information is legal and in no way controversial. It is part of the game. The problem lies in the transition from what is legally acceptable to criminal behaviour. The dividing line can be related to what methods are used, whether ethical guidelines are broken in order to acquire information and whether the measures violate the laws of a country or not. Different nations have different legal systems; an activity can be viewed as criminal in one country, but be legal in another.

Industrial espionage is relatively cheap compared to investments in advanced research and development. According to an estimate by the FBI, industrial espionage in 1992–1993 cost more than 120 billion U.S. dollars in lost contracts and R&D expenses. The number of lost jobs was assessed to be 6 million.[144] Later estimates have figures of more than 200 billion U.S. dollars annually, in the United States alone. In Canada the cost of illegal information collection is estimated at more than 12 billion dollars per year.[145]

In 2009 the information security company Symantec conducted a survey in order to analyse the amount of information stolen, and the cost of it. A total of 2,100 companies in 27 countries participated in the study. The result showed that all the companies that participated had lost important information; in 92 percent of the cases it had led to great costs. Each information theft cost an average of nearly 2 million dollars.[146]

---

144 Robert Lyle, Radio Liberty/Radio Free Europe, 10 February 1999. Quoted in "the European Parliament Report 2001 on the existence of global system for the interception of private a commercial communications (ECHELON interception system)" (2001/2098(INI)) RR\445 698EN.doc.
145 "Defectors say China running 1,000 spies in Canada". *CBC News*.
146 Danielsson, L. (2010) "Räkna med spioner i företaget". *Computer Sweden*, 23 March 2010.

Espionage can be sanctioned at national level and/or be part of an individual company's strategy to gain competitive advantages. In some cases third parties are used for the actual information collection, for instance a criminal organisation or a company. One of the most serious threats comes from insiders. They may be planted in an organisation by a security service or the like. An insider can also be an employee who has been recruited to conduct a specific task. His or her motivation may be financial or personal, such as dissatisfaction with the work situation and a desire to cause the company harm. People can also be bought, bribed, blackmailed or forced to hand over vital information. A player can manipulate a person into handing over secret information without understanding the consequences his/her actions, who is actually behind the operation or why it is conducted. Referring to ideology, patriotic, ethical and/or sentimental reasons are viable tools for recruitment.

After the end of the Cold War there was a drastic increase in industrial espionage. One reason may be that many security services were forced to change concentration after the fall of the Berlin Wall and adjusted to a new situation. Today it is not necessarily military capacity that is the decisive means of pressure between regions and states; economic strength is just as important. Industrial espionage is also very lucrative with great payback on invested capital. It is a tempting activity. If an operation is discovered it is often legally very hard to tie the actions of individuals to a security service from a specific country, for example. In some cases the interests of a regime coincide with those of a company, in other cases a company may act as a proxy and a cover for an operation. It is important to stress that industrial espionage is in no way limited to enemies; it can also be conducted between friendly nations and companies that are not seen as competitors.

Advanced technology can be acquired through industrial espionage, technology that can be used to build up industrial capacity and speed up development of products. A country that is technologically less developed can use a structured and conscious long-term collection campaign for sensitive information, which can provide technological leaps that make the country catch up with, and sometimes even overtake, its opponents.

For individual companies, sensitive information can give an insight into the plans of a competitor and how it will act in certain situations. That way counter-measures can be worked out at strategic and operational levels. Information can also be used for military purposes, e.g. to modernise armed forces. In a security policy context, such as negotiations between states, strategic information can be decisive for the outcome. It can lead to a better negotiation position and the opportunity to exert pressure. For states with export restrictions, industrial espionage is a viable method to circumvent the sanctions and acquire the technologies they need. For countries like Iran and North Korea, which according to the UN arms embargo are banned from purchasing strategic products on the open market, industrial espionage may be a solution.

However, industrial espionage is not the only way of acquiring important information. One effective method is the indirect one; make companies voluntarily give away vital information, knowhow and capital. China, especially, has developed a successful strategy. The same is true of India, when it comes to outsourcing of IT operations and services from the West. Both nations have learned from the Japanese industrial policy in the 1950s and 1960s.

In China's case it is about attracting foreign capital through false pretences about its immense market; how important it is for Western companies to be represented there in order to maintain their global competitiveness. Additionally, large and advanced companies that want to set up business in China are required by the government to open joint research institutes in the country. There are presently more than a thousand research laboratories, within a number of fields, financed by foreign companies.

Offers are "sweetened", and Chinese industrial policy makers are very well aware of and prepared to adjust to the Western capitalist (often very short-sighted) philosophy, aiming at fast profits in order to please shareholders. During the establishment phase in a country like China, many companies pay no or very little considerations to security policy, ethical or environmental aspects, other than rhetorically in their annual financial reports.

The bait, apart from the huge domestic market, is cheap and highly qualified labour. The short-term profits can be high for companies that invest in the country; but the long-term costs can be even higher, unless they are aware of the risks and prepared to act accordingly, which unfortunately many Western companies are not. Generally speaking it has been quite hard to get a good return on investments.

There are a number of examples of how Western industries have invested in advanced production facilities, product development and research, but the outcome has not been as expected. The car industry is a striking example. Companies such as Volkswagen have indirectly helped China develop its own capability and capacity to manufacture cars efficiently. In the long run, this strategy may result in a backlash. The first Chinese-produced, and from a cost perspective competitive, cars have started to be exported to the West. The quality is still inferior when it comes to safety, for example, but this will change as there are more robust designs, similar to Japanese products in the 1960s and 1970s.

General Motors has accused the Chinese state-owned car manufacturer Chery Automobile Co., Ltd. of industrial espionage, with reference to the Chery QQ being a copy of GM's Chevrolet Spark. Legally this has been hard to prove. According to GM, two employees of Chinese extraction stole information about the company's investments in hybrid technology, worth 40 million dollars.[147] Even the design of Volkswagen models is seen as having been copied by Chinese car manufacturers.

An area of increasing interest for companies, and one that may lead to severe information drainage, is the trend of storing information in "clouds". Instead of investing in costly systems, servers and databases, companies outsource the storage to "secure cloud computing services" firms. Another term for this is "virtualisation", which can be done at different levels of the IT systems.[148]

The main purpose of outsourcing IT functions is cost-cutting. The companies do not have to invest in expensive equipment. But just as when it comes to outsourcing production capacity, IT operations or research and development to a third party, the consequences of virtualisation involves security problems that have to be carefully investigated from case to case.[149]

"Cloud computing" means a physical and geographical reallocation of resources. For a parent company it may be difficult to check their security company's routines on a daily basis; who has access to what kind of information and under what conditions. The risk of unintentional leaks to a third party is quite high, especially if there is a large geographical distance between the parent company and the outsourced activity. There may also be cultural differences between countries and companies in their views on sensitive information. Good agreements can to some extent regulate the security issues. Another matter that has to be considered is what happens in case the security company is purchased by another rival.

---

147 Garthwaite, J. (2010) "Hybrid Espionage! GM takes couple to court over tech secrets". *Erat2tech*. 23 July 2010.
148 Virtualization could be made on three levels: on servers, in data storage and on client level.
149 "US Airforce Seeks Cyber Security in a Cloud". *Defence IQ*, 26 April 2010.

The number of official cases where companies and research institutes have lost vital information, and ended up in court, is relatively low. The reason is that it is often in the company's interest not to announce that it has been the victim of information theft, since this could hurt their reputation and affect customer relations.

There are some historical examples of espionage with industrial implications. One is German physicist Klaus Fuchs, who worked as a spy for the Soviet military intelligence service, the GRU. During the period 1947–49 he handed over information on how to construct hydrogen bombs to the Russians. That way the Soviet nuclear arms programme could be reduced by several years. Another example of suspected industrial espionage is the development of the Soviet supersonic passenger plane Tupolev-144, sometimes called "Konkordski". As the name indicates there is a striking resemblance to the French-British Concorde. The first flight of the Soviet aircraft took place at the end of 1968, two months before Concorde.[150] However, on the whole, Tupolev-144 was not a commercial success. Its history was lined with various technical problems. At the 1973 Paris Air Show one aircraft crashed, killing 14 people.

A very alarming affair, with great implications for security policy regionally and globally, is Pakistani nuclear scientist Abdul Qadeer Khan's activities. After university studies in Karachi and research studies in Belgium and Germany in the 1970s Khan started working in Holland for Urenco, a joint British-Dutch-German consortium specialised in developing methods for uranium enrichment. In 1976 he returned to Pakistan and became head of the country's nuclear programme. Two years earlier, in 1974, India, Pakistan's main rival, had successfully set off a nuclear bomb. This created a lot of concern in the Pakistani regime. Khan and his research team were tasked with developing a Muslim bomb that could be used as a deterrent. The first charge detonated in 1998.[151]

After having moved from Holland, Khan was sentenced in absentia to four years in prison for stealing blueprints to centrifuges, vital technology in the enrichment of uranium for nuclear bombs. He has also been accused of having shared technology that can be used to enrich nuclear fuel with controversial states like Libya, under Muammar Gaddafi, Iran and North Korea. The last two are seen as "rogue states" by a majority of the international community.

---

150 "The Russian Concorde Tupolev TU-14". *Gizmo highway.*
151 Bowcott, O. (2009) "Abdul Qadeer Khan. Pakistani nuclear scientist accused of industrial espionage". *Guardian.co.uk*, 6 February 2009.

The British counterintelligence and security agency MI5 primarily singles out China and Russia as very active in industrial espionage.[152] According to MI5 there is some uncertainty as to how widespread their activities are, but especially the Chinese strategy is assessed to be very focused, with a global agenda concentrated on high-tech industries. It is important to remember that illegal collection of information is in no way limited to these two players; a number of countries and companies are involved in these kinds of activities, to a greater or lesser extent.

An article written by J. Hill describes the Chinese industrial espionage threat.[153] In the U.S. there is a lot of concern that the country is the subject of illicit intelligence collection on a massive scale. This is also true of Western Europe. In Hill's article Chen Yonglin, first secretary at the Chinese consulate in Sydney, says there is a network with more than 1,000 spies and informers in Australia alone. Areas of interest include industry and finance, especially the IT and telecom sectors, as well as sensors, radars, biotechnology and energy systems.

The article also mentions that Canada is the Western country with the most Chinese spies. The reason for this may be both the great contingent of Chinese living in the country and the short distance to the U.S., which is viewed as a primary target. Europe is also beset. The Belgian security service, for example, has said that several hundred Chinese spies are working at different levels in European industry.[154]

The Chinese telecommunications company Huawei has been repeatedly accused of having copied Western companies' equipment, and then cut prices on networks and equipment. According to information on the Internet, Huawei representatives have visited subcontractors with blueprints of Ericsson base stations and asked them to manufacture more or less identical stations. The quality of the source is uncertain and the information has not been confirmed.

There are also rumours that Huawei has copied network switches from Ericsson and router technology from Cisco. The allegations have been fiercely denied by Huawei. The alleged military connection between the company and the Chinese armed forces, and the rumours of industrial es-

---

152 "Chinese Espionage: Britain's MI5 reports epidemic in spying". *Examiner.com*.
153 Hill, J. (2005) "Detection s Reveal Extent of China's espionage Operations". *Jane's Intelligence Review*, vol. 17, 1 November 2005.
154 Cole, M. (2009) "Friendship is no bar to espionage". *MOSS. MacArthur Centre for Security Studies*, 1 November 2009.

pionage and piracy, has made it difficult for Huawei to establish itself in the U.S. In 2003 Cisco, in an American court, accused Huawei of having copied data codes used in routers.[155] According to the court records, the Chinese company had even copied Cisco's serial numbers, in order to make it easier for potential customers to switch from Cisco's equipment to Huawei's. In 2004 Huawei employee Yi Bin Zu was caught photographing sensitive equipment from Fujitsu during an industrial fair in Chicago. Compromising information was discovered on his USB memory card.[156]

Indian security services are very critical of Huawei's involvement in the country. In the spring of 2009 Huawei, together with its rival Ericsson, won a huge contract worth five billion dollars from Indian telecommunications operator BSNL.[157] Together, the two companies were going to deliver mobile networks. Huawei was going to be responsible for infrastructure in the southern part of the country, while Ericsson was responsible for the northern and eastern parts. The contract was frowned upon by the Indian security services and armed forces. The reason for the concern was that the equipment could be used for illicit information collection. The company would also be given the opportunity to assume control of information infrastructure of national importance to India. In the spring of 2010 the agreement was cancelled, a great loss of prestige to the Chinese. The event should also be seen in the light of the frosty relations between the two countries, among other things due to the 1962 border conflict. It can also be seen as an expression of increasing rivalry between two economically growing superpowers in Asia and globally.

Information on advanced telecommunications systems is a field of great interest to the Russians. In 2003, a 46-year-old engineer was sentenced to eight years in prison for industrial espionage against Ericsson. The man, who had worked on product development, handed over secret information to an intelligence officer at the Russian embassy in Stockholm, Sweden. At the same time, an accomplice was sentenced to four years.[158]

In Germany a case of industrial espionage, with links to the Russian intelligence services, was revealed in 1999; it is believed to have been going on for several years. The investigation showed that two employees from the

---

155 Simmons, C. (2006) "The Huawei Way". *Newsweek*, 16 January 2006.
156 Simmons, C. (2006) "Huawei in Spying Flap". *Newsweek*, 24 June 2004.
157 "Indisk konflikt kan gynna Ericsson". *Dagens Industri*, 15 May 2009. "BSNL gets nod to award to Huawei". *The India Times*.
158 Cervenka, A. (2010) "spionjägare". *Svenska Dagbladet*, 18 July 2010.

DASA defence company had been passing on information on weapons systems to Russian agents.[159]

In 1988 two Israelis employed at the American nuclear weapons programme in Los Alamos penetrated the central computer and stole information. After the operation the hackers fled to Israel, where one of them was prosecuted. All links to the Israeli intelligence services have been publicly denied.[160]

Even the Scandinavian airline SAS has conducted illegal information collection on its competitors. In spring 2010 the company was fined more than 20 million dollars by a Norwegian court of appeal for industrial espionage on its rival Norwegian. SAS had illicitly acquired information on its rival from the Amadeus booking system.[161]

Cyber espionage

Cyber espionage is a generic term for a number of activities. The purpose is the illegal collection of information through the Internet, networks or individual computers by using different penetration technologies and/or planting malicious code in a target's computers and networks. The objective is to get access to sensitive information, but also to gain control of a computer system without the owner's permission or knowledge.[162] By using deliberately planted trojans the computers can be taken over.

The Internet, satellite links and wireless communications provide new ways of conducting espionage. Through signals intelligence, voice calls and data traffic can be recorded, registered and monitored. Sensitive data can be intercepted, deciphered and analysed. It is also possible to follow traffic intensity in a geographical area; if e.g. a certain country's embassy increases its communications, this indicates something is going on.

Nowadays, cyber espionage is an important part of the global economic competition. Monitoring an opponent's computer systems and capacity is

---

159 "Annerkennungen zur sicherheitslage der Deutschen Wirtschaft". ASW, Bonn. Published in the European Parliament Report 2001.
160 Förster, A. (2001) "Maulwurfe in Nadelstreissen". Published in the European Parliament Report 2001.
161 Meltzer, J. (2010) "SAS har dömts för industrispionage i Norge". *Affärsvärlden*, 17 March 2010.
162 CSR Report for the Congress. "Botnets, Cybercrime, and Cyberterrorism: Vulnerabilities and Policy Issues for Congress". Order Code RL32114. Updated 29 January 2008.

seen as an inevitable part of national defence. The exchange, that is the value of the collected information in relation to the cost of collecting it, is very good. This means that these kinds of activities will keep increasing.

The U.S. is very worried about cyber espionage. The threat is especially severe for countries that build their industrial strength on innovations and immaterial rights. Naturally, this also applies to other countries, both in Europe and elsewhere. In 2008 sensitive data worth an estimated 150 billion dollars were stolen through cyber espionage in the U.S.[163] The amount includes not only industrial espionage, but also trade secrets, intellectual property, such as patents, and diplomatic relations.

A 2008 Congress report on cyber threats quotes sources in the security services, where it is said that intelligence services from as many as 140 countries are trying to penetrate the systems of American companies and the government.[164] John Tkacik, in an article from 2008, says that a leading member of the U.S. counter espionage community, Dr John Brenner, has singled out China and Russia as prominent players in cyber espionage.[165] The Chinese activities are said to be the single biggest risk when it comes to theft of American technology and know-how. Cyber penetration of networks and computers is, according to the report, the most effective tool for the Chinese security services, and it is used against targets in both the U.S. and among its allies.

Officials in Australia have also stated they have problems with Chinese cyber espionage. The Australian national security service ASIO, along with the Defence Signals Directorate, accuses China of being behind a number of attacks on the Internet and attempts at cyber espionage against the country.[166] Here is one example. In an article in *The Wall Street Journal*, 21 April 2009, there were reports of security problems surrounding the multibillion dollar investment in the new Joint Strike Fighter. According to the article, the project had been compromised by hackers, allegedly from China. Large amounts of data, terabytes, were stolen.[167]

In the spring of 2007 the German federal security service, BfV, went public with information that the IT systems of the Ministry of Foreign Af-

---

163 Ackeman, R. (2009) "Threats Imperil the Entire U.S. Infrastructure". *Signal*, July 2009.
164 Ackerman, R. (2009).
165 Tkacic, J. (2008) "Trojan dragon: China's Cyber Threats". *Heritage*, 8 February 2008.
166 Roy, B. (2009) "China's Silent Warfare". *South Asia Analysis*, paper No. 3147, 13 April 2009.
167 Gorman, S., Cole, A., Dreazen, Y. (2009) "Computer Spies Breach Fighter-Jet Project". *New York Times Journal*, 21 April 2009.

fairs and the Ministry of Research and Development had been attacked by hackers located in the Lanzhou, Guangzhou and Beijing regions.[168] According to *The Economist* more than 160 gigabytes of information was sent from the system before it was stopped.[169] Trojans were embedded in Microsoft Word documents and PowerPoint files.[170] The German weekly *Der Spiegel* states that 60 percent of all cyber attacks against Germany originate from China. Chinese hacker attacks against German infrastructure are conducted more or less on a daily basis.

The United States Secretary of Defense Robert Gates in June 2007 admitted that unclassified computers connected to the NIPRNet in the Pentagon had been penetrated by Chinese hackers.[171] The work to restore the system took several weeks. A possible target may have been the servers of the Department of Defence.[172] (U.S. government agencies and defence contractors were attacked more than 80,000 times by various antagonists in 2007.) According to a 2006 communiqué from the United States Department of Defense, the Pentagon's network, which is a part of the Global Information Grid, is allegedly scanned more than three million times every day.[173]

But the U.S. has also been accused of conducting cyber espionage. There are suspicions that the country is developing advanced technical capacity for strategic signals intelligence that can be used for illegal information collection. A system that has been discussed widely is the American "Echelon" network. In a 2001 EU report, the American signals intelligence system was analysed, its capacity to monitor private and commercial data traffic in Europe and whether it is used for industrial espionage.[174]

The report establishes that such an intelligence system exists. The network operates at global level, under the command of the National Security Agency (NSA), with a number of monitoring stations in several countries. Echelon covers countries such as the U.S., Great Britain, Canada, Australia and

---

168 Nicoll, A., et al (2007). "Chinas cyber attacks. Casting a wider intelligence net". *IISS Strategic Comments*, vol. 13 issue 07, September 2007.
169 *The Economist*. Vol. 384, No. 8545, 8 September 2007.
170 Schecter, E. (2008) "U.S. struggles to break ahead of rivals in network security race". *C4ISR Journal*, 1 March, 2008.
171 Nicoll, A. (2007).
172 Dong-A, I. (2007) "China Wants Dominance in Cyber Space". *ROK Daily*, 21 September 2007.
173 Carr, J. (2010) "Inside Cyber Warfare. Mapping the Cyber Underworld", *O'Reilly media Inc.*, December 2009.
174 European Parliament Report 2001.

New Zeeland. The report assesses the risk of domestic European Internet traffic being monitored to be low; on the other hand, traffic transmitted via the U.S. may be tapped. The same applies to radio traffic conducted at national basis. That kind of monitoring – that is, of radio channels, not cables – requires a relative physical closeness to the object of interest. The traffic that above all is intercepted by Echelon is satellite communications, according to the report. At the same time it is pointed out that that traffic is marginal compared to the flow of data in cables and in cellular systems. When it comes to the issue of whether the monitoring system is used for industrial espionage or not, the report is vague. There is no clear-cut answer in the report.

An important part of the NSA's system for interception is the multi-billion dollar site *Utah Data Center* outside the small town of Bluffdale, Utah. When fully functional in 2013 the facility, connected to listening posts, will increase the possibility to intercept, analyse and decipher all kinds of communications, not only personal but also financial movements and transactions.

It is clear that there has been American industrial espionage, though not necessarily via Echelon. In 1994, for instance, negotiations were in progress between the European industrial consortium Airbus and a Saudi national airline about the purchase of aircraft. In this context the CIA intercepted fax traffic and telephone calls. The information was forwarded to Airbus competitors Boeing and McDonnell Douglas. The American companies won the contract, worth six billion dollars.[175]

In 1996, people linked to the CIA are said to have hacked the computers of the Japanese Ministry of Trade. The purpose was to steal information that could be used in negotiations on import quotas of American cars to Japan.[176]

Russia has very good capacity for strategic signals intelligence and surveillance. According to law *No. 40–FfZ*, passed by the Duma and signed by President Boris Yeltsin in 1995, the FSB (which took over after the KGB) has the legal right to monitor telephone lines, open mail and survey the Internet. The law makes it possible for the FSB to conduct intelligence operations in Russia and abroad in order to secure the country's interests.[177]

---

175 "Antennen gedreht, Wirtschaftwoche". No 46, 9 November 2000. Published in the European Parliament Report 2001.
176 Schutze, von A. (1998) "Wirtschaftspionage: Was macht eigentliche die Konkurrennz?". No 1, 1998. Published in the European Parliament Report 2001.
177 "On Organs of the Federal Security Service in the Russian Federation". Russian Federation Law No, 40-ZFZ. Adopted by the State Duma, 22 February 1995. Signed by the president Boris Yeltsin and dated 3 April 1995.

The system is called SORM II.[178] This means that all communications transmitted via operators such as Rostelekom, Transtelekom and Elektrotelekom are passed on to the FSB.[179] In 2012 the Duma voted for new and harder restrictions of Internet usage for Russians.

Just as the American NSA, the Russian foreign intelligence service (the SVR) is believed to have signals intelligence capacity. Together with the military intelligence service (the GRU), the SVR manages monitoring stations in Yemen and on Socotra in the Gulf of Aden, off the coast of Somalia.[180]

It is unclear whether there are any links between the Russian security services and nationalist hacker groups when it comes to cyber espionage. There are very few official sources confirming the information. The editor of the Russian hacker magazine *Khaker*, Sergei Pokrovsky, has confirmed that the FSB hires hackers for both international and domestic espionage.[181] In a report, Alexander Osipovich mentions that there may be links between hackers and the security service.[182] In an interview in *Pravda* in 1996, the head of the GRU, General Feodor Ladygin, admitted they have the capacity to hack computer networks in order to acquire sensitive information.[183]

Although there is very limited official information that confirms links between the Russian security services and hacker groups, there is a historic case from the Cold War called *Operation Cuckoo's Egg*.[184] The background was the following. In 1985, the KGB hired East German hacker Markus Hess to hack the computers of U.S. defence authorities. Hess was prosecuted, accused of having attacked as many as 400 American computers connected to the civilian Arpanet and the military MILNET, both forerunners of today's Internet. Hess was found guilty of espionage and sentenced to prison in 1990.

---

178 SORM is an Russian acronym for "System of Operation Research Measures" similar to the FBI's Carnivore system and the British "Government Technical Assistance Centre (GTAC)"; see Leijonhielm, J., Hedenskog, J., Knoph, J., Oldberg, I., Unge, W., Vendil, C. (2000) "Rysk militär förmåga i ett tioårsperspektiv. En förnyad bedömning 2000". Användarrapport FOA.
179 InfoSecurity (2009) "Grey Goose 2 Ties Kremlin More Closely to Georgia Cyber-attacks". *InfoSecurity Magazine*, 27 December 2009.
180 *InfoSecurity Magazine* (2009).
181 "Cyber warfare". Internet article.
182 Osipovich, A (2007) "Inside a Hacker school". *Foreign Policy*; Nov/Dec 2007 issue 163.
183 Carr, J. (2010).
184 Stoll, C. (1990) "The Cuckoo's Egg: Tracking a Spy Through the Maze of Computer Espionage". *Pocket Books*.

The General Staff's third department of the People's Liberation Army – 3PLA – is in charge of strategic signals intelligence in China. This includes capacities such as information collection, deciphering and analysis of electronic signals from satellites, fixed and wireless networks within as well as outside the country. 3PLA has a well-defined capacity to stop unwanted communications. The department is also responsible for cryptanalysis. It is assumed that 3PLA develops cyber warfare tools such as viruses and trojans as well as hacking methodology. There are indications that 3PLA mans a number of signals intelligence sites abroad, for instance in Laos, Burma, Thailand, Cuba and on several islands in the South China Sea. According to information in open sources, Chinese embassies perform signals intelligence in foreign countries.

The Ministry of Industry and Information Technology, MII, is responsible for controlling information transmitted between private persons, companies and organisations over telecommunications systems and computer networks. The MII monitoring apparatus is based on "web police". More than 20,000 people are allegedly involved in surveying Internet usage alone. According to the official news agency Xinhua, China's three leading telecom operators monitor the contents of all SMS traffic distributed in their networks. Western companies such as Google and America Online have assisted Chinese authorities by installing blockers and filters in search engines in order to censor information.

The main task of the web police is censorship. But in order to censor content, someone has to read, analyse and assess how sensitive the information is. Interesting information is probably forwarded to the system of those in power. The capacity of the MII, the security service (the Ministry of Public Security, MPS) and other law enforcement authorities and military organisations to intercept information is assessed to be very good.

Surveying the Internet is done both manually and automatically. Every day the web police send lists of words, key phrases, subjects and events that have to be erased. Words and sentences that are considered offensive are deleted. People who are seen as having acted inappropriately on the Internet are controlled and – if the case is seen as serious – prosecuted.

As computerisation has become more and more widespread, there is a large amount of sensitive information stored in networks and on employees' computers that can be accessed quite easily. The digitisation has made it possible to convert, package and distribute information in new ways. This creates a number of security problems. USB memory sticks are an especially sensitive area. In a very short period of time, a person can download a large amount of data, take it with him and hand it over to a third party, without it being noticed. A subtle way is to deliberately "forget" a USB memory stick contaminated with malware after a business meeting, hoping someone will be curious and put it in his or her computer.

One example that illustrates the risks of this is *Operation Buckshot Yankee*. In 2008 the U.S. Department of Defense discovered a breach in a secret computer network. It began with a USB memory stick that infected an American laptop in the Middle East. The malware, a trojan, was planted by a foreign intelligence service and uploaded itself to a network controlled by the U.S. Central Command. The trojan spread, without being discovered, to both open and secret computers, and opened up "loopholes" where sensitive data was transmitted to servers under foreign control.

This computer network operation is the single most serious incident to have been discovered so far. During the time it was active, thousands of files were forwarded from U.S. and allied networks and industrial partners to a foreign country. This includes blueprints of secret equipment, military plans and surveillance data.[185] The incident led to a rude awakening for the Pentagon. Cyber espionage against the country has become a prioritised problem to solve, lifted to highest military and political level. It is still not known what state was behind the operation. Officially, no country has been singled out.

Another security problem is employees losing or forgetting USB flash drives, computers and cell phones in public places. This can be unintentional, through carelessness or thoughtlessness. But it can also be due to theft, where someone is deliberately trying to get hold of the information stored on the devices.

New ways of spreading trojans are constantly being developed. One innovative method is the following. Instead of infecting attachments in e-mails with malware that are sent to persons of interest, a human intelligence operation is conducted. The target and his behaviour and habits on the Internet are surveyed. If it turns out that the victim regularly visits certain websites, at work or in his spare time, they are infected with malicious code. When the user visits the website the next time, a trojan is automatically downloaded to his computer. It starts gathering information from his network and computer and sends it on to a third party.

---

185   Lynn, W. (2010) "Defending a new Domain". *Foreign Affairs. Pepperdine University. School of Public Policy*, September 2010.

Another creative way of deceiving persons of interest is to send e-mails that appear to come from the company's internal IT support. The e-mail message says that the computer has been infected or taken over, and a large amount of information has been sent from the system. In order to fix the problem, all the victim has to do is click on the attachment. And in the attachment trojans are, of course, embedded.

The Chinese Ghost Net

28 March 2009 *The New York Times* published an article that was given a lot of attention. A very large network used for cyber espionage had been discovered that probably emanated from China. The background to the event was the following. For a couple of years, the Tibetan Government in Exile had suspected that their computer network had been infiltrated by the Chinese government. In order to find out whether this was the case they asked the organisation *Information Warfare Monitor* (IWM)[186] to investigate the matter. IWM is loose international organisation consisting of a number of computer-skilled people from the SecDev Group, a Canadian consulting firm, and security experts from the University of Toronto and the Cambridge Computer Laboratory in Britain. For ten months they gathered information, analysed traffic patterns and got a picture of a global espionage network called *GhostNet*.[187]

The cyber espionage network had not only been directed against Tibetans in exile; its scope was global. Indian embassy computer systems had been compromised, just as embassies in a number of countries such as South Korea, Indonesia, Romania, Cyprus, Malta, Thailand, Taiwan, Portugal, Germany and Pakistan. The foreign ministries of Iran, Bangladesh, Lithuania, Indonesia, the Philippines and other countries had also been infected. For a short time NATO computers were allegedly also under surveillance. IWM came to the conclusion that the operation was directed against a total of 103 countries, and 1,295 computers had reportedly been infected by trojans. Details of the methods used to infect the networks are

---

186 Http://www.infowar-monitor.net.
187 Markoff, J. (2009) "Vast Spy System Loots Computers in 103 Countries". *The New York Times*, 28 March 2009.

described in a 56-page report, published by the organisation and available for download on the Internet.

In brief, the following method was used. First, e-mails with attachments infected with trojans were sent to specific, selected targets. When the recipients opened their e-mails, malicious code infected their computers. The trojans were in contact with servers in China. Through a specific command, some of the infected computers were ordered to download and install another, even more advanced, trojan designated *Gh0st Rat*. The Rat code enables the infected computers to be taken over and controlled in real time from the servers. This made it possible to, for instance, activate functions such as cameras and recordings by remote control.

Almost 350 of the more than 1,000 computers that were compromised have been traced to South East Asia, more specifically to government offices in diplomacy and finance. In Europe and on the American continent, Belgium and Canada were above all targeted, but computers were also taken over in countries such as Finland, Germany, France and Spain.

The IMW experts have not been able to confirm that it is the Chinese government that is responsible for the operation. But in a report from Cambridge it was mentioned that the Chinese state probably participated at least in one intrusion into the offices of the Dalai Lama. It is possible that the architects behind GhostNet are Chinese nationalists with financial objectives, or people linked to the Chinese intelligence services.

In the investigation of servers and IP addresses, IMW found indications that the Gh0st Rat trojan originates from the island of Hainan in the South China Sea. That is an interesting piece of information, since the island houses not only a major naval base, but also the Lingshui signals intelligence office, which is part of the Chinese army's strategic activity.[188]

Ghostnet infected not only Chinese rivals, but also nations it co-operates with, such as Pakistan and Iran. The cause behind this is a moot point. The Chinese may have an interest in finding out what their "friends" are doing. It may also be an act of deception; if the operation is discovered, which it now has been, the perpetrators can say that they would never do this to their friends, and consequently have nothing to do with the operation, i.e. some kind of deniability. According to Indian China expert Bashkar Roy, it is not surprising that China also monitors its friends, it is a silent strategy that fits

---

188 "Chinese hackers using ghost network to control embassy computers". *The Times*, 29 March 2009.

the nation's ambitions to achieve global leadership.[189] The statement should be viewed in the light of India's and China's complex relation as growing superpowers.

Canada is one of the countries with most infected computers. This is interesting in view of the fact that Canadian authorities have identified problems with Chinese industrial espionage. The same applies to Belgium, where they have identified insiders and advanced industrial espionage with origins in China.

Because of the development of computers and the Internet, "classic" industrial espionage converges with cyber espionage. Digitisation makes organised theft of information easier. The ability to penetrate and gather sensitive information from afar, but also gain control over networks and computers and remote control them, is very worrying. Methods will gradually be refined, improved and adjusted to different kinds of targets.

The profitability of the spy business, combined with the difficulties prosecuting the individuals and organisations behind these kinds of operation, opens the door not only for security services but also for new players like cyber crime organisations. Lowered barriers and the possibility to generate large revenues make cyber espionage very attractive. It is a trend that will just keep getting stronger.

---

189   Roy, B (2009) "China's Silent warfare". *South Asia Analysis*, paper No. 3147, 13 April 2009.

Cyber crime

Along with drugs smuggling and arms trafficking, cyber crime is one of the most profitable and fastest growing criminal fields in the world. Crimes over the Internet involve a number of activities, such as spreading malware and spyware, spamming, blackmail through threats of DDoS attacks and defacement of websites, password thefts, phishing, pharming, carding and hacktivism. There are strong links between illegal information theft, i.e. cyber espionage, and cyber crime, where criminals both steal information and sell it to punters on the black market. Pharmaceuticals and drugs smuggling, as well as trafficking and other sex crimes over the Internet, are also part of the business.

The first time computer crime was identified as a phenomenon was in the 1970s. People with good technical skills joined traditional white collar criminals in order to commit crimes against companies.[190] Gradually these kinds of insider crimes spread from disloyal employees to people outside the system, such as hackers and crackers. Their aims were not necessarily financial; just as often the perpetrators wanted to show off their skills to other hackers. The goal was status; the motto "hack for fame".

The first viruses appeared in the 1980s. The purpose was to a large extent to destroy information in systems – not to steal data. At this time, Internet usage was still limited and immature, so malicious code was spread between computers through infected software on floppy discs that passed from hand to hand. This changed in the 1990s, when the Internet started becoming popular. Viruses could be spread more effectively via the Internet; they also became more intelligent and advanced. A new characteristic was the ability to self-replicate. The new viruses that were introduced were relatively few, with names such as My Doom, Melissa and so on. In those days, the people behind the viruses only wanted recognition in hacker circles. Gradually, the focus changed from status to money, from "hack for fame" to "hack for fortune". As a consequence of this development, parts of the hacker community

---

190   Chies, R., Ducci, S, Ciappi, S. (2009) "Profiling Hackers. The Science of Criminal Profiling as Applied to the World of Hacking". *CRC Press Taylor & Francis Group*, NW, USA.

started to merge with organised crime. The number of viruses introduced every year is no longer in the hundreds, but in the tens of thousands, and they keep getting more advanced.

Today cyber crime is a fully developed, very lucrative and innovative business. The opportunity to earn a lot of money attracts individuals, groups and organisations with little or no scruples from all corners of the world. This kind of criminality is directly related to the development in IT, with an ever increasing number of activities on the Internet. Where there are people – regardless of whether it is in the physical or the virtual world – there are also criminals. Crime follows people's behaviour. The increase can also be related to the generally low level of knowledge, awareness and protection skills among large parts of the population when it comes to threats and risks on the Internet. The lack of security awareness also constitutes a big problem for companies and governments.

There are a lot of different players. Some hacker groups are suspected of having links to specific regimes and their security services; others to organised crime and the mafia. Other kinds of players include loosely organised groups and individuals who act independently without any links to a third party. The objective of the activities is mainly to make money illegally. If a security service is behind such crimes, political and economic motives can be the main reason, but be hidden behind ordinary criminality. Targets for these activities include financial institutions, companies and individuals.

The whole criminal scale is covered, from minor attacks during a short period of time to extensive and well-organised operations. Co-operation across borders is common; one perpetrator may be based in one country planning and co-ordinating the operation, while the executors are scattered all over the world. Criminals are becoming increasingly professional and innovative in their behaviour. The crime scene is global; activities performed across national borders. The risk of detection and prosecution by law enforcement agencies has so far been low, which has made the business attractive to many players. Since 2000 law enforcement agencies have become increasingly active and skilled in the field, and cross-border co-operation is improving. That way, more players can be brought to justice.

How serious is the threat? According to the FBI, cyber crime is number three on the scale – after terrorism and espionage.[191] The black economy of hackers and data thieves is a multi-billion-dollar business in Europe alone.

---

191   Krebs, B. (2007) "security Fix. Calculating the Costs of Cyber Crime". *The Washington Post*, 27 September 2007.

The majority of the players producing code can be found in Eastern Europe, Russia and Asia. The United States has singled out Russia and China as particularly aggressive in this regard.[192] These two countries are seen as an unregulated area and a safe haven for various kinds of cyber crime.[193] There are a number of prominent and infamous hacker groups in South America, Africa and the Middle East, and in the West, of course. The U.S. is by no means a safe zone, rather the opposite. Some of the most dedicated and notorious cyber criminals reside in North America.

An article published in *The Washington Post* in 2007 discusses the difficulties in counting the costs of crimes over the Internet for businesses and individuals.[194] Both tools and methods are lacking. But there are some estimates. The FBI, for instance, calculated the costs for American firms in 2005 to 67 billion U.S. dollars.[195] According to Shawn Henry, assistant director for the FBI's Cyber Division, the costs for identity theft and fraud of Americans alone is estimated at 26 billion U.S. dollars per year. These kinds of crimes are usually conducted through hacking. Chris Hoofnagle at the Berkeley Centre for Law and Technology says that cyber crime in the U.S. alone has an annual turnover of tens of billions of dollars.[196] The computer security company McAfee estimates the global cost of cyber crime to be on the same level as cyber espionage.[197] But cyber espionage is also criminal.

In order to get a perspective on the sums they can be related to the cost drugs smuggling and the arms trade. According to information from the UN, the narcotics trade had an annual global turnover of 321 billion U.S. dollars in 2003.[198] Since then it has probably increased. Global expenditures for weapons were 1,500 billion U.S. dollars in 2009, according to the organisation Global Issues.[199] The Stockholm International Peace Research Institute (SIPRI) estimated global arms exports in 2007 to have been worth

---

192 Keggler, J. (2008) "Taking the Fight to the Net". *Armada International*, vol. 32, issue 2 (April/May),
193 Ackerman, R. (2009) "Threats Imperil the Entire U.S. Infrastructure. From the military to the economy, the country is open to vast damage". *SIGNAL, AFCEA International Journal*, July 2009.
194 Krebs, B. (2007).
195 FBI Computer Crime Survey 2005.
196 Hoofnagle, C. (2007) "Identity Theft: Making the Known Unknowns Known". *Harvard Journal of Law and Technology*, vol. 21, 2007 (autumn).
197 Mills, E. (2009) "Cybercrime cost firms $1 trillion globally". *CNET News*, 28 January 2009.
198 Pollard, N. (2005) "UN report puts world's illicit drug trade at estimated $321". *The Boston Globe Journal*, 30 September 2005.
199 Shah, A. (2009) "World Military Spending". *Global Issues*, 7 July 2010.

50 billion U.S. dollars.[200] An educated assumption is thus that the cost of cyber crime worldwide is in the hundreds of billions of dollars annually.

What is the situation in my native country, Sweden, for instance? There is no definite information. According to the police, the equivalent of 500 million U.S. dollars are lost annually in investment scams, many of them over the Internet.[201] Every year, thousands of people are victims of fraud and other kinds of deceit. But this is just a minor part of cyber crime. The total costs are probably much higher. And the situation is more or less the same in many modern countries all over the world.

One of the most serious activities in this field is theft of credit card numbers and financial information through hacking. In January 2009, for example, Heartland Payment Systems, the sixth largest credit card processor in the U.S., announced it had been the victim of one of the biggest computer breaches in history in 2008, involving close to 100 million credit card numbers.[202]

This is not the only time something like this has happened. At Christmas 2008, hackers succeeded in getting into RBS WorldPay's computer system for electronic transactions. This U.S. service company provides the payment system for the Royal Bank of Scotland. On this occasion, account information for more than 1.5 million customers was stolen, along with social security numbers and other personal information.[203] Swedish banks are by no means protected from this kind of activities. In May 2007, more than 10,000 credit card numbers were frozen, after they were stolen from transaction company Alphacash.[204] Two years earlier, there was a major breach of the Swedish payment system. 24,000 credit card numbers were stolen, which makes it one of the biggest Internet frauds in Sweden. All the big banks were victims, so were banks in other parts of Europe.[205] It's very hard to get any information about the affair; the banks are silent.

---

200 SIPRI. "The Financial value of global arms trade". 11 October 2011.
201 "Smarta högutbildade offer för aktiebluff". *Dagens PS*. 21 September 2010.
202 Claburn, T. (2009) "Heartland Payment System Hit By Data Security Breach". *Information Week*, 20 January 2009.
203 Adhikari, R. (2008) "RBS WorldPay Data Breach Hits 1.5 Million". *Internetnews.com*, 24 December 2008.
204 Goldberg, D. (2007) "Handelsbanken drar tillbaka kortnummer". *Computer Sweden*, 7 April 2007.
205 Goldberg, D., Larsson, L. (2007) "Banker tiger om stulna kortnummer". *Computer Sweden*, 20 June 2007.

However, banks and financial institutes are not the only targets. A new area of fraud is stealing accounts for online games such as World of Warcraft and Second Life. The logic of this kind of role-playing games (RPGs) is based on creating digital persons – avatars. If a user is skilled and gets more experience points by playing the game well, the value of the avatar increases. An avatar can be sold at online auction sites like eBay for real money, up to hundreds of dollars per character. Payment is done via credit cards. In connection with the transactions, criminals can hack accounts. Tens of millions of people play RPGs over the Internet every day. There is a lot of money in circulation. South Korea, along with Japan, is usually described as the country where people are most involved in computer games. It is therefore no surprise that South Korean game servers have allegedly been hacked and information stolen. In the middle of April 2011 there was a massive hacker assault against Sony and its Internet-based application PlayStation. The online service was forced to shut down. Almost 100 million accounts were hijacked. The cost for the two week suspension of the online service – along with the loss of confidence in Sony as a trademark – amounted to almost 20 million U.S. dollars.

A classic area for criminals is violence, or the threat of it, in order to blackmail companies, organisations and individuals. On the Internet there can be blackmail of this kind via threats of defacement – i.e. changing the appearance of and information on corporate websites – or DDoS attacks. This is especially serious for companies that rely on the Internet for their business, such Internet banks, e-commerce and online companies. On an individual level hackers can steal vital and personal information, such as sensitive e-mail conversations, and threaten to spread the material to third parties, unless he is paid not to do so.

Nordic media company Canal Digital was subjected to a blackmail attempt by hackers in 2008.[206] The criminals allegedly succeeded in manipulating the digital codes of their TV cards. The cost of changing half a million customers' TV cards was almost 3 million U.S. dollars.

In 2006 Russian authorities arrested a group of criminals that threatened a number of British online betting companies with DDoS attacks. The group allegedly succeeded in blackmailing more than four million dollars from Internet casinos and betting firms in Great Britain.[207] One of the com-

---

206 "Hackare ville förhandla om med Canal Digital". *Dagens Nyheter*, 19 October 2010.
207 "Online Russian blackmail gang jailed for extorting 4 million USD from gambling websites". *Sophos*, 5 October 2006.

panies that was being blackmailed, Canbet Sport Betting, refused to pay 10,000 dollars. During one of the year's most important sports events, the Breeders' Cup horse races, the company website was inaccessible. Every day the betting firm lost the equivalent of 200,000 dollars. Thanks to a unique co-operation between Russian authorities, the UK's National High Tech Crime Unit, Interpol and the FBI, the activities of the group were unwound. Three people were sentenced to eight years each in prison; they were also sentenced to pay a lot in damages to their victims. In the subsequent investigation it was revealed that for six months the group had made fifty similar blackmail attempts in thirty different countries, with various degrees of success.

The event described above is quite small and limited, but it shows the opportunities for cross-border crime. Security firms have expressed their concern that enormous botnets will be used for DDoS attacks on a massive scale, for blackmail purposes or as part of a bigger operation. According to some estimates a botnet of one million compromised computers could generate enough traffic to "sink" most Fortune 500 corporations for some time. A botnet with 10 million computers could theoretically paralyse the network infrastructure in many Western countries.[208] It does not require a great deal of fantasy to imagine the consequences if someone were to threaten to saturate the networks of the markets, such as Wall Street or the Frankfurt Stock Exchange.

Are there such large botnets? According to some security experts, yes. The *Conficker* Internet worm, for example, is specially designed to attack computers with the Windows operating system, and is estimated to control more than five million computers, including authorities, companies and home computers in more than 200 countries. The computer power in this botnet exceeds the capacity of some of the largest computer centres in the world.[209] According to the website of Shadowserver Foundation, a non-governmental organisation which follows cyber crime on the Internet, Conficker has infected more systems than any other malicious code so far.

Conficker was first discovered in November 2008. It is unclear who is behind the code – whether it is a criminal gang or an intelligence or security service in order to monitor and survey its opponent's computers. What security experts have found out from studying the design of the code is that

---

208   Carr, J. (2010).
209   Markoff, J. (2009) "Defying Experts, Rogue Computer Code Still Lurks". *The New York Times*, 28 August 2009.

the first version contained a stop function that prevented computers with Ukrainian keyboards from being infected. This information may give a hint of its origin. It might also be part of a deception operation, and somebody else is actually behind it. Since then, new versions have been discovered. Whether Conficker is the big threat that some people say is something that has to be investigated further.

A relatively common criminal activity is fraud through *phishing*. Phishing is a method to get hold of sensitive information, such as passwords and bank account numbers. A common course of action is the following. A criminal group designs a website very similar to another website that belongs to a well-known bank or some other financial organisation, using the same name, information and logo. Customers that use bank services over the Internet are misled to visit the false website and type in their passwords and account numbers in the belief that they are on the legitimate website. That way the criminal group gets access to the login information that is required get access to the customer's accounts.

In 2006 Scandinavian bank Nordea was subjected to an advanced phishing attack. Using e-mails, swindlers tried to mislead net clients to log on to a false website. The event was repeated six months later.

A sophisticated kind of phishing is the following. A criminal group presents itself as an Internet security company and sends out information to potential customers with an offer to subscribe to their security programs. The website of the false security firm looks very trustworthy, using the same design as established companies in the business. If a customer chooses to buy their products, payment is done over the Internet on their website. When this has been done, a program is automatically installed on the customer's computer, and security patches are continuously being downloaded in a way that makes it look as if they come from a legitimate company.

There are three levels to this fraud. Firstly, the security program does not protect the computer. The customer pays for something that obviously does not work. Secondly, the software contains malware that is installed which both opens up the system and makes it possible to steal credit card information. Last – but not the least – a relation is established between the parties, in which the swindlers can milk the user for a long time without him/her knowing that he/she is the victim of fraud.

An elaboration of phishing is *pharming*. This means that a visitor to a legitimate website, such as an online bank, is automatically redirected to a false site, where spyware is installed on the computer. This is done even if the right web address (URL) has been used. One difference between the methods is that phishing is individual, one visitor at a time, while pharming has a lot of victims.

One annoying part of everyday activities on the Internet is *spam*, the purpose of which is to advertise via e-mail. For the client it is a very cheap method to market products and services. Each e-mail can be copied millions of times at no extra cost. As long as a fraction of a percent of Internet users believes the message and buys the products the client makes money. The problem is dual; these unwanted messages hijack a lot of network capacity and there is a risk that some of the e-mails and their attachments are infected by malicious code. The Internet is contaminated by spam. For Internet service providers who run the networks, the phenomenon results in extra costs. They have to adjust to the amount of data transmitted and received, and thus invest due to an increased level of traffic. Companies can use filters to block unwanted traffic, but this also requires investments.

There are different kinds of spam. Junk mail can be divided into unsolicited commercial e-mails (UCE[210]) and unsolicited bulk e-mails (UBE[211]). The cost of spam for companies and individuals in terms of disruptions, saturation of broadband capacity, reduced work opportunities and so on are assessed to be very big. The yearly cost of handling spam in Sweden alone is estimated to be somewhere between 300 and 800 million dollars, and this does not include hardware or administrative costs.[212]

According to the Spamhaus Project, an organisation that keeps statistics of incidents on the Internet, 80 percent of all spam to users in North America and Europe comes from a core of a couple hundred professional spam gangsters. The ten worst spammers include six from Russia and Ukraine; the others are based in India, China, Estonia and Australia. The countries that generate most spam are the U.S., China and Russia.

Russian hacker Nikolai McColo, founder of the McColo criminal network, was a notorious spammer. During a period up until his death in a car accident in Moscow in September 2007, more than 75 percent of the entire world's total spam was sent from computers compromised through his network. After his demise the amount of spam decreased by 67 percent overnight.[213]

---

210 UCE – unsolicited commercial e-mail.
211 UBE – unsolicited bulk e-mail.
212 "spam kostar miljarder". *IDG*, 12 March 2004.
213 Carr, J. (2010).

Theft of bank and credit card information, *skimming*, is a growing problem. This is the typical insider job. Many people fall victim when visiting restaurants, shops and withdrawing cash from ATMs. Hijacking credit cards has become an important way to make money for criminals. The yearly turnover is substantial, causing big problems for the victims.

There are several ways of stealing information. The simplest method is to photocopy receipts. A more advanced kind of fraud is to attach "skimmers" to ATMs that can steal hundreds of customers' credit card numbers. The perpetrators often add a small camera that registers the personal identification numbers.

A new and even more refined way (instead of manipulating card readers or adding skimming equipment to ordinary card terminals and ATMs) is using false card readers that are not connected to the card holders' banks. It works in the following way: In a transaction, the card terminal copies the secret account information but does not send it on to the bank, i.e. the purchase is never registered. Instead, the information is saved in the false card reader. Based on the information that was copied, new cards are made. A lot of money can be withdrawn using the card, even weeks after the first theft was made. It is very hard for the victim to know when and where the hit was done.

A big problem for the gaming, music and software industry is *cracking*. Cracking means that a special program code, or script, is used to unlock or circumvent copyright protected software, such as the source code of operating systems. There are two kinds of cracking, computer and software cracking. The first is about destroying systems, the second about breaching protected software. Companies put a lot of effort into preventing breaches of their products. Developing software script is difficult and demanding. This means that the best crackers are often insiders, working as software testers at IT companies, with access to secret information.[214]

Market places

Just as the open and legal economy, criminals need working market places that allow them to launder money and trade or sell stolen property. The

---

214   Chies, R. et al. (2009).

need for areas that are safe from the eyes of law enforcement agencies is of course vital to cyber criminals too, with the distinguishing characteristic that what is sold is digital information, not physical products. The price of the information follows the standard economic logic based on supply and demand. Generally speaking, cyber criminals are very dedicated and innovative. They learn new technologies and solutions fast. In some cases they have better skills than the ones investigating their activities.

There are a number of secret websites and fencing servers on the Internet, so-called *underground economy servers*, which are used by criminal organisations and hackers. These websites market opportunities to commit fraud. Examples of what is offered include malware, such as Trojans, lists of cracked credit card numbers, with CVV2 codes, personal identification numbers and lists of e-mail addresses.[215] There are a number of Internet forums where hackers can exchange information on suitable targets to attack and the methods to do it. In chat rooms, insiders working on cracking protective codes may sell beta versions of new software, including computer games, movies and future releases of operating systems. Cracked films can often be bought on the Internet before they have been commercially released.

Symantec's 2007 annual report on cyber threats said a great deal of these fencing servers can be found in Sweden. The countries that host the most servers are the U.S., with 51 percent of all the discovered servers, followed by Sweden with 15 percent and Canada with 7 percent. According to Symantec's 2009 report, the life expectancy of the servers is less than six months. The reason for the short life expectancy is that those in charge of the websites want to avoid detection by constantly moving the transactions to other sites and locations. When one site closes down another one opens up, but under a different name and in some other corner of the Internet. This makes it hard for law enforcement agencies to follow what is happening.

An illegal cyber transaction can be done in the following way. First, a criminal organisation conducts a hacker attack and steals credit card information. The group does not use the stolen card numbers, instead they contact a "cashier" who acts as a middleman, linked to a fencing server. The cashier publishes information on an illegal website about the opportunity to buy cracked credit cards. Visitors who want to conduct credit card fraud log on to the website and check out what is on offer. If the visitor finds something interesting he contacts the cashier and discusses the price. When the parties

---

215 Symantec. "Internet Security Threat Report 2007".

have agreed, the visitor buys some of the stolen credit card numbers for a certain price. When the deal is concluded, the cashier transfers the money to the criminal group, minus his cut. The visitor uses the cards to buy products and services online. The costs are charged to the owners of the real cards. In the best cases that owner is compensated by the financial institution that issued the card, when the fraud has been discovered. To cover their expenses, the costs are shared by all the customers in the bank.

Buying cracked credit cards is not expensive. According to Symantec, the market price is somewhere between 10 cents and 25 dollars, depending on the kind of card, how current the information is and the credit rating of the card holder. This means that a criminal can buy credit card numbers, including personal identification numbers, for less than the cost of a lunch, while the value of an individual account may be thousands of dollars. Between July 1, 2007 and June 30, 2008 the security company estimates the global cost of stolen credit card information and fraud was 5.3 billion dollars.[216]

Cyber crime in Russia

As mentioned in the first chapter of this book, a lot of malicious activities in cyberspace originate from Russia. This includes a wide range of activities, from spreading malware and spam, to cyber crime, cyber espionage and hacktivism directed against its adversaries. In this context Russia, together with China, is seen as an unregulated area and a safe haven for the development and dissemination of malicious code worldwide.

Thanks to its high educational standards in the natural sciences, mathematics and physics, there are a great number of skilled and IT trained people in the country. For instance, in the Moscow area alone about 250,000 people are employed in the information and communications technology (ICT) sector.[217] For young people it is quite hard to get a job. Many well-educated young people apply for jobs abroad. According to estimates, more than 70,000 Russians work in the American IT industry. At the same time, the software industry in Russia is growing, especially firms in the information security sector. Russian programmers are considered among the best in the world

---

216 "New Symantec Report Reveals Booming Underground Economy". *Information Systems Security*, 24 November 2008.
217 IT & Software opportunities in Moscow. Moscow Investment Gateway. Source: Moscow Government; Goskomstat; *Statistics on Russian Education*; Watson Wyatt.

and are often used by Western companies such as Microsoft, IBM, Google etc. There are several universities in Russia that teach computer sciences and network security with high international standards and good renown.

There are several elements behind the forces driving the development of malicious cyber activities. Unemployment among the young generations is one problem. The 2009 unemployment rate was approximately 8–9%, and the social security system is inadequate.[218] Corruption in Russia is huge and covers all levels of society. There is a big and growing gap in living standards between people in the big cities and the rural parts, something that creates social tensions. Laws protecting property, not least intellectual property, are also weak, as is the court system.

Many citizens do not see cyber crime – such as phishing, identity and credit card theft, Internet fraud, hacking into banking accounts, website defacements in order to blackmail companies and organisations – as major threats to society, especially not when compared to other kinds of crime. Moreover, IT infrastructure is relatively well-developed, especially in the major cities.

Most activities are directed against foreign, not Russian, commercial websites run by banks and financial institutions. If cyber crime is not aimed at Russian interests and local targets, law enforcement agencies generally do not know much, nor do they have much motivation to investigate and take legal actions. If foreign companies fail to protect their own systems, it is their problem, not something for Russian authorities to handle. Another reason is the possibility to be anonymous on the Internet and to conceal digital traces; this makes it hard for law enforcement agencies to discover, trace and catch potential criminal cyber activists. This occupies a great deal of resources that could be better used against more prioritised kinds of crimes.

An increasing problem is Russian spammers who steal personal information and money from bank accounts. There are special websites run by criminal groups where one can buy lists of stolen credit card numbers that can be used for fraud. Information on how to hack into commercial systems is available on Internet forums. One well-known hacker website is the Khaker (Russian for hacker) website, xakep.ru. Khaker also publishes a magazine with the same title that can be bought online. There are also "hacker schools" that teach basic skills in how to crack computers and network systems. One well-known hacker school has had around 10,000

---

218   Russia Employment Rate. Trading economics. *Global Economics Research*. (2009).

students since it was founded in 1996.[219] The courses are advertised in the media.

There is evidence that Russian organised crime syndicates are involved in cyber crimes, too. The modus operandi when it comes to co-ordination, sophistication and the choice of specific targets indicates that cyber attacks are committed by a well-financed, organised and experienced group (or groups) of criminals.[220] Any links to the Russian security services are very hard to prove. There is very little information and the sources are to a large extent highly uncertain.

One infamous group of cyber criminals is the Russian Business Network (RBN). The RBN has been involved in various aspects of cyber crime, such as phishing, distribution of malware and malicious code, botnets, DDoS attacks and even child pornography.[221] The group provides a more or less complete infrastructure for crime on the Internet.

Its history can be traced back to early 1996. In 2002, the group became more organised and structured, and its activities increased. For instance, the RBN was accused of attacking the U.S. Department of Defense and the Russian Department of the Treasury in 2003, though this has not been officially proved.[222] Up until 2007 the group acted as an Internet Service Provider (ISP), leasing servers for all kinds of crimes.[223] Between early 2006 and November 2007, when the RBN served as an ISP, it was linked to 60% of all cyber crime, according to information security company Verisign.[224] It is a business worth millions of dollars.

In 2007 the RBN's IP addresses and domains were blocked and blacklisted by the information security community, this forced them to move their domain servers to China and Taiwan. It is uncertain whether the group is still active and operational.[225] According to information on blogs, the network was

---

219 Osipovich, A. (2007) "Inside a Hacker School". *Foreign Policy*, issue 163 (November/December).
220 Flook, K. (2009) "Russia and the Cyber Threat". *Critical Threats*,13 May 2009.
221 Flook, K. (2009).
222 Verisign (2008) "The Russian Business Network: Rise and fall of criminal ISP". *IDefence Security Report*, 8 March 2008.
223 "Russian Business Network (RBN)". *Blogspot*, October 2007.
224 Verisign (2008) "The Russian Business Network: Rise and fall of criminal ISP". *IDefence Security Report*, 8 March 2008.
225 Shactman, N. (2008) "U.S. Embassy in Russian Hackers' Crosshairs?". *Wired*, 12 August 2008.

involved in the Georgian cyber conflict in 2008.[226] Others claim it no longer exists. Due to the lucrative nature of the business, the RBN probably continues its activities, but in different forms and under another name.

The RBN acts on an international arena with links to other criminal networks. One example is California-based Attrivo, also known as Intercage. This company is specialised in renting networks for spam and other hostile actions, some of them linked to the RBN. Attrivo controls hundreds of proxy servers,[227] divided into 15 networks in a number of countries. Especially two networks have been used for hostile purposes: UKrTeleGroup, which routes traffic to Ukraine, and HostFresh, which transmits traffic through Hong Kong and China. Under pressure from the authorities, Attrivo was forced to shut down on 22 September 2008. Attrivo's most important customer was ESTDomain, a company based in Tartu, Estonia, that specialises in domain addresses used for criminal activities. In 2008 the manager, Vladimir Tsatsin, was sentenced to three years in prison for credit card theft, document fraud and money laundering.[228]

Chinese criminality

Just as Russia, China is frequently accused by Western countries of harbouring cyber crime organisations; accusations that are vehemently denied by the Chinese authorities. At the same time, the political leadership admits that cyber crime is a growing problem, even for China. For instance, Minister of Public Security Meng Jianzhu in December 2009 declared that the police should act "with force" against the increasing criminality on the Internet. In June 2010, the Ministry of Information published a white paper on the growing threat from cyber crime. In 2010 the Ministry of Public Security[229] conducted raids against several hacker groups, 450 hackers were prosecuted. The ministry has also stated that crime on the Internet increased by 80 per-

---

226 Macquaid, J. (2008) "The RBN Operatives Who Attacked Georgia Secure Home Network". *Blogspot*, 18 August 2008.
227 A proxy server is a server filtering request – which is forwarded to other servers making the process anonymous.
228 Carr, J. (2010).
229 MPS – Ministry of Public Security.

cent between 2009 and 2010; it added that the attacks mainly came from China and were mainly directed against Chinese targets.[230]

The main reason for the increase in activities by Chinese hacker groups is that they are very lucrative. A difference between Chinese and Western groups is that many of the Chinese are driven by patriotism, which is not necessarily the case for groups in the West. The boundary between politically and criminally motivated hacktivism is thus not quite clear. Generally speaking, the Chinese authorities frown on hacker activities directed at Chinese interests, irrespective of whether for money or political purposes. On the other hand, if the targets of the attacks are foreign it is not seen as serious, and the authorities are often reluctant to act.

More advanced Chinese hacker groups use another logic than normal among many Western hackers. They do not provide free advice on the Internet on how to launch attacks, which is common behaviour in order to achieve status in hacker circles, i.e. "hack for fame". Another interesting difference is that Chinese hackers normally do not use Facebook or other social networks for co-ordination and exchange of information; instead they prefer the instant messaging service QQ.

According to a newsletter on the Internet, hacker groups such as China Eagle Union (CEU) have contacts with the Chinese mafia, the triads.[231] But since CEU can be characterised as nationalist, it probably would not act against the interests of the Chinese state. CEU could rather be a tool for the authorities to gain some control over the crime syndicates. Whether this is the case or not, should be regarded as speculation. Officially, Chinese authorities try to suppress all kinds of crimes.

One trend described in blogs is that Chinese hackers have transformed from nationalism to commercialism and into cyber crime.[232] Big money can be earned by stealing and selling credit card numbers and identities, hacking online game sites and developing and providing viruses and trojans to third parties on an industrial scale. There are indications that many small and mid-sized companies have been forced to pay ransoms to hacker groups for "protection" in order to secure their websites and businesses, such as e-commerce, on the Internet from attacks. However, this is not only

---

230 Noonan, S. (2010) "China and its Double-edged Cyber-sword". *Stratfor*, 9 December 2010.
231 "Some Experts Question Activities of China's Patriotic Hackers". *World News Connection*, 7 October 2007.
232 The Dark Visitor. 4 April 2008. Internet article.

a problem for the rest of the world; to a large extent it is the same thing for Chinese consumers.

A representative of the National People's Congress, Chen Wanzhi, says the authorities have to get better control of the criminal subculture that has developed on the dark side of the Internet. In Hunan province, police arrested a hacker who had specialised in stealing banking accounts on the Internet. He had allegedly gained control over more than a thousand bank accounts and stolen more than 400,000 yuan, almost 50,000 U.S. dollars. Chen says cyber crime has grown into an industry divided into different business areas. Some develop trojans and viruses, others spread malicious code. Other business areas include stealing information from banks, laundering and depositing money earned from criminal activities.

One of the most profitable business areas is to sell "loopholes", i.e. security flaws in corporate websites. Another is to lease botnets with compromised slave computers for dissemination of viruses. The price of a trojan varies depending on how advanced it is, from 0.1 to more than 1,000 yuan. According to information from the Chinese cyber incident organisation CNCERT,[233] reported in a blog, there are several botnets in China with more than 100,000 overtaken "zombie" computers each. These zombies can in turn be leased to hackers for purposes of blackmail, earning them millions of yuan. In order to prevent this, the Ministry of Public Security has closed down more than a hundred websites were hackers could download programs and tools for different kinds of cyberattacks.[234]

Another profitable line of business is to hack computer game servers, take over avatars and sell them on auction sites such as eBay. According to one report from CNNIC,[235] reported in a blog, the number of online gamers in China in 2008 was more than 40 million. Profits from the online gaming industry were estimated to be more than 30 billion yuan in 2010. This is definitely an interesting target for less scrupulous hackers.

There are indications that, apart from domestic cyber crime, unholy alliances are being formed between Chinese and foreign organisations, especially the Russian mafia. According to unconfirmed information joint coordinated groups have attacked websites in order to steal bank account numbers.

---

233   CNCERT – China CERT
234   Noonan, S. (2010).
235   CNNIC – China Internet Network Information Center.

According to the well-known information security expert Eugene Kaspersky, Chinese cyber criminals are developing very advanced multi-vector backdoor trojans, and trojans adapted to steal data from online games.[236] This shows that there are people with very good skills in designing malicious code in the Chinese region. Kaspersky also says that when it comes to cyber crime, China is now No. 1 in the world – well ahead of the Russian mafia.

Through reversed engineering, several information security companies and IT incident organisations have observed that certain kinds of trojans originate from mainland China. Through binary investigations of the text in the trojans' code, Chinese characters have been discovered. Although there are strong indications, it is difficult to single out a specific country with absolute certainty. However, in this case the indications are convincing and they point at China; that does not mean that the Chinese security services are involved. It is also important to stress that piracy of operating systems and other software is big business and common in Asia. The software often has security problems due to lack of regular patching that fixes security problems.

The Business Software Alliance estimates that 79 percent of all software sold in China in 2009 was pirated.[237] This in turn opens up for a number of vulnerabilities, such as infections from hostile code. The lack of information security is a big problem in China, which probably has authorities worried.

Cross-border criminality

Compared to Eastern European, Russian and Chinese cyber criminals, North American ones are not as organised and structured, but more ad hoc. Groups are put together for an operation and subsequently disbanded. Some of the most dedicated and cunning people in this criminal business are based in the U.S. Initial contacts often take place on Internet Relay Chat (IRC) channels and online forums, where there are discussions about the possibilities to attack different targets and the best ways of doing so. In order to restrict outsiders following chats, they sometimes take place in

---

236 Weinberg, N. (2007) "Kasperskys on cybercrime: Don't blame the Russian maffia and why we need anti-anti-anti virus software". *Network World*, 2 January 2007.
237 Noonan, S. (2010).

closed peer-to-peer networks, such as TOR. Communication is often enciphered. There are presently very big, global IRC networks. The opportunity for criminals from all corners of the world to co-operate is very good. There is no need for meetings in real life, instead planning and realisation of an operation can be carried out digitally.

One of the most notorious American cyber criminals is Albert Gonzales, who was captured in May 2008. He has been accused of, together with two Russian assistants, being behind the largest theft of credit card information in history, so far. Between October 2006 and May 2008 more than 130 million account numbers were stolen from Heartland Payment Systems. 7-Eleven and Hannaford Bros supermarkets were also attacked by the troika. Gonzales has also been accused of stealing 40 million credit card numbers from TJX Companies, a subsidiary of the clothing company TJ Maxx, at an estimated cost of 200 million dollars. Even Barnes & Nobles, the bookstore chain, belongs to his victims.[238]

American law enforcement agencies were not amused. In 2009 he was sentenced to twenty years in prison. The term indicates how seriously this kind of criminal activity is viewed. It is quite ironic that the Secret Service, in order to prevent cyber crime, paid Gonzales 75,000 dollars per year to act as an informer, during the time of his criminal activities.

Another example of the development toward cross-border criminality is the group of 60 people that the FBI arrested for fraud in October 2010. This was a Ukrainian-controlled network with branches in the U.S. and Great Britain. The group had tried to empty bank accounts worth the equivalent of 220 million dollars in small and medium-sized companies in several countries. A common denominator for all these companies was very bad security in general, and especially poor information security.

In order to get into the systems the group used the Zeus trojan, which is written to steal banking information by logging key strokes. That way passwords and login information for bank account numbers can be stolen. The trojan is spread via downloads and phishing attacks. It was first discovered in 2008 and can be bought on fencing servers for between 70 and 300–400 dollars, depending on the version.[239] The Ukrainian network probably purchased the malicious code from an illegal server. The Zeus trojan is seen as one of the big security problems in the 2010s. According

---

238 "Cybercrime Kingpin Indicted". *The New New Internet*, 18 August 2009.
239 Stevens, K., Jackson, D. (2010) "Zeus Banking Trojan Report". *Secure Works*, 11 March 2010.

to one assessment it is probably the second most infectious in the world, after the Conficker worm.

One problem that keeps getting bigger with the growth of the Internet is the classic Nigerian scam. This is a global criminal activity, based in Nigeria, with links to Holland and England, among other countries. On the Internet there is a group that calls itself *Fraud 419* which has been active since 1992. "419" is the official Nigerian legal code for this kind of scam. Nigeria letters can be characterised as an "advance fee bluff". The fraud takes place through text messages and e-mails. What seems like personally written letters are transmitted to a very large number of people. The typical message is that the sender needs help transferring a large amount of money, often several million dollars, under a false name. For a fee, the receiver can have some of the money, provided that he/she deposits some money into a bank account belonging to the swindler.

The astonishing thing is that this scam works, over and over again, despite all the warnings in the media. According to Stephanie Nolen at *The Globe and Mail*, an experienced scam artist can make thousands of dollars from each letter that is answered. According to one estimate, Fraud 419 gets one or two replies for every thousand mails.[240] That is a pretty good reply rate, especially when bearing in mind that it is more or less free to send e-mails on the Internet.

The thing that makes people reply is the usual – greed. There are examples of people having deposited tens of millions of dollars into false accounts. The results of this kind of fraud are often tragic; victims have even committed suicide when they discovered they had been scammed. People have even been killed when trying to get their money back. One example is Norwegian businessman Kjetil Moe, who was beaten to death by the Nigerian mafia in Johannesburg, South Africa, in September 1999 after a settlement of 6,000 dollars.[241]

The structure of the organisation is complicated. The network is divided into different layers, depending on the number of operational cells. In each cell there is a person who is designated to act as a bag man for the money that the cell earns. The money that comes into the network is passed on to

---

240 Stephanie, N. (2005) "Nigerian e-mail scammers feeding on greed, gullibility". *The Globe and Mail*, 5 December 2005.
241 "Moe funnet död". *Verdens gang*. Dagbladet.

the leaders higher up in the hierarchy.[242] There are indications that this kind of scam is sanctioned by the top levels of Nigerian society.

There are also other kinds of fraud. In the autumn of 2010 a huge swindle was revealed that originated from Holland but was directed at a number of European countries. Almost 500 Swedes are believed to have been cheated out of 60 million U.S. dollars by false stockbrokers.

The victims – often educated, intelligent and level-headed people with a good financial situation – were contacted by salesmen over the telephone and offered to buy unlisted shares with seemingly good potential.[243] The salesmen were polite but obstinate in their attempts to convince their victims to do business. The deposit was usually 10,000 dollars or more. In order to create a fashionable front, the scammers used stockbrokers Witter Walwyn Overseas Ltd. as a cover. Potential clients had nice-looking brochures sent to their homes. Buyers could follow their investments and trades on a daily basis on false websites connected to the stockbrokerage firm. How the scammers selected their victims is not known. There probably was some sort of survey of possible victims via the Internet and phone books.

Afterwards, when the scam has been revealed, it is often hard for law enforcement agencies and prosecutors to investigate the chain of events, since very few of the victims are prepared to admit publicly that they have been cheated. They feel ashamed and stupid. The chance of getting back the stolen money is very small. The scammers cover their tracks and move on to new targets.

What conclusions can be drawn from these kinds of scams? One is that greed is a strong driving force. Another is that the number of people that can be fooled is, if not endless, at least countless. If an offer sounds too good to be true, it probably is. It is just as true in the real world as it is in the virtual one. The Internet is an invitation to different kinds of criminal activities *en masse* and the annual turnover from Internet crimes is huge.

A big, profitable market combined with a low risk of discovery will make this kind of activity continue to grow. Law enforcement agencies in many countries are trying to co-ordinate their resources nationally and co-operate internationally. This is essential, since we are talking about cross-border, global criminality. There are some examples of successful co-

---

242 "Nigeriabrev. 419 Scam". *Antarktis media*, 2002-2004.
243 "Smarta högutbildade offer för aktiebluff". *Dagens PS*, 21 September 2010.

operation; co-ordinated efforts have resulted in smashed criminal networks with branches in several countries.

One problem is that laws and regulations are one step behind the technical development. They are not adapted to cyber crime. The legal system also differs between countries. In some areas property laws – material as well as intellectual – are very weak. It creates a scope for this kind of crime. The difficulties bringing cyber criminals to justice is also due to the fact that many of them are very cunning, quick and innovative. Where there is money there are always people who will try get a piece of the cake, through legal or illegal methods. Cyberspace is an unregulated market place where crime flourishes. These activities can also be seen as training in skills that can be used in destructive contexts – for instance hacktivism and other kinds of cyber warfare.

Aggression between hackers

Hacktivism – or politically, religiously, ethnically or ideologically motivated hacking of opponents' networks and servers – is a growing problem. Example of antagonistic activities include "defacement" of websites, where information is changed or removed, and distributed denial of service, DDoS, attacks that overload websites making them inaccessible.

Hacktivists are by no means a uniform group, but consist of individuals and groups from all corners of the world. Some of them have only rudimentary skills in computing, but are driven by an intense belief in their cause. Others are smart, skilled and very experienced people, organised in groups and dedicated to their task. In some cases state players with large resources can be suspected of wanting a third party to act on their behalf, in what is usually called cyber war by proxy.

A conflict in cyberspace usually has its origins in the real world, where hacktivism is used as a force multiplier to convey messages and show one's strength to the adversary. Using the Internet for activism is an option when other means may not be available or appropriate. It is a very useful and cheap tool for conveying messages.

The origin of the activism may be a perceived offense against something of great idealistic or religious value. Individuals or groups of people feel wronged for some reason and want to present their view or express their indignation. Hacktivism can also aim at supporting one party in a conflict in progress between nations or ethnic groups. In some cases economics and power are the causes, but they are hidden behind a fog of political, religious or nationalist messages.

When a conflict is initiated in cyberspace, a situation arises where many people are willing to help one of the parties to attack the opponent. Through the globalisation of the Internet, people from different parts of the world can participate in an operation. Temporary alliances are formed between disparate groups that normally do not have anything to do with one another. However, in a certain situation they are united and act coherently towards a common goal. On the opposite side, hackers unite to protect their own networks and carry out counterattacks.

The conflict is kindled on various Internet forums, with accusations and outright attacks. Quite often, misleading information is spread on websites, Facebook and in chat channels. The purpose is to inflame. This can lead to a dangerous escalation of the conflict between the hacker groups. Increasingly heavy artillery is used, more and more targets are attacked, many of them do not even have anything to do with the actual conflict. In order to facilitate large-scale attacks, information is published on websites describing how visitors can go about attacking the opponents. Temporary sites are set up with instructions and links to software that can be downloaded and used in DDoS attack in order to block the activities of the other side.

One example of conflicts from the real world that have spilled over into the virtual world is the relentless Internet confrontations between Palestinians and Israelis, and their respective supporters. Attacks have gone back and forth, with shifting intensity, and struck both sides, and third parties that have little or nothing to do with the conflict. Hacktivist activities have been especially rampant in connection with the Palestinian Intifadas.[244] Attacks are often psychological, directed against the other side, but also the international community. It is about inspiring sympathy for the cause, antipathy to the opposite side. The tactic is based on striking against the main rival's propaganda channels, such as media sites and the political leadership's websites. But attacks are also directed against systems vital to society, such as banks and financial institutes.

When the second Intifada started 28 September 2000, pro-Palestinian hackers e.g. succeeded in attacking web sites hosted by the Bank of Israel, the stock exchange in Tel Aviv and the Israeli army. The Israeli economy depends on electronic commerce; stocks closed down eight percent.[245] During the Intifada period, pro-Palestinian groups managed to attack more than five times the number of sites that Israeli groups did. In an analysis of the event it is hard to see whether the main cause of the drop on the stock exchange was the cyber operation or the war itself. However, it is quite clear that the digital activities did influence the market to some degree.

"Operation Cast Lead", the Israeli invasion of the Gaza strip in December 2008, caused new tensions on the Internet. One of the main reasons for

---

244 Intifada is an Arabic term for "uproar". Several intifadas have been proclaimed; the first one was in 1987 and the second started 28 September 2000 when the Israeli president Ariel Sharon attended the mosque in Jerusalem. Source: Wikipedia.
245 Atwan, A.B. (2006).

the operation was the recurring rocket attacks against Israeli civilian targets by political-religious Palestinian organisation Hamas. One interesting difference compared to previous conflicts between the parties was that the Israelis seem to have involved hackers. According to one source, people from the Israeli Defence Forces conducted psychological operations against Hamas TV station Al-Aqsa. Hackers succeeded in getting the station to broadcast a short movie showing pictures of Hamas leaders losing their lives, with subtitles in Arabic saying: "Time is slipping away".[246]

The month-long operation caused a great number of civilians casualties, mostly on the Palestinian side. The exact numbers are disputed and differ depending on what side comments. Many people, especially in the Middle East, reacted strongly against what was seen as Israeli outrage. In order to support the Palestinian cause, Turkish hackers, together with groups from other countries, attacked thousands of websites in Israel and pro-Israeli countries and organisations, among them the U.S. Army's and NATO's web sites.[247] The method used was quite simple, but politically effective, such as defacement of opponents' websites where information was replaced with an anti-Israeli manifesto.

One example of a target was English-language portal Ynetnews.com, one of the leading newspapers in Israel. The Morocco-based group *Team Evil* succeeded in redirecting traffic from Ynetnews and other Israeli websites to a site with a protest against the conflict. Other defacements were directed against Israeli Cargo Air Lines and political party Kadima. The latter was conducted by a group from Algeria, *DZ Team*. The party website showed photos of funerals of Israeli soldiers, with a text in Arabic announcing even more Israelis would die.[248] Israeli Defence Minister Ehud Barak's website was also attacked, probably by an Iranian group called *Ashiyaneh Security Team*. The message in English said: "ISRAEL. You have killed more than 800 innocent civilians in Gaza. Do you think that you won't pay for this? Stop the war. If you don't we will continue hacking your important sites."

According to IT security expert Jeffrey Carr, who among other things started "Project Grey Goose", the ambition of which it is to study damaging activities on the Internet, most of the hackers in the Gaza conflict seem to

---

246 Carr, J. (2010).
247 Raza, M.A. (2009) "Thousands of Websites hacked by Muslims hackers to Protest Gaza Attacks". 16 January 2009. http://propakistani.com.
248 Carr, J. (2010).

have been of Moroccan, Algerian, Saudi, Turkish and Palestinian extraction. This does not imply that the attacks came from these countries; they may have been carried out from other locations. There was no joint or coordinated group behind the attacks; they seem to have been spontaneous.

Confrontation Islam – the West

There are sign that Islamist hacker groups based in different countries and regions are organising themselves and directing their activities against servers and websites in the West, especially in Israel. There are a number of infamous groups such as the *Iranian Cyber Army* and *Osama Bin Laden Crew*. In a similar way, individuals and hacker groups in Europe, the U.S. and other parts of the world are organising themselves to attack Islamist websites and hackers. One example is *Internet Haganah*, an American-based Zionist group, probably with the capacity to shut down Islamist websites.

A recent case that got a lot of attention in the media is the controversy surrounding the Danish Muhammad cartoons, which led to a Middle East boycott of dairy products from Denmark, assaults on consulates and flag burnings. The event also generated cyber attacks against websites. Groups of Islamist hackers, for instance *Gangs of pro-Muslim hackers* attacked Danish websites with the suffix "dk" more than six hundred times during one week.[249] Danish hackers and hacker groups associated to them in their turn tried to protect their domains and counterattack. Similarly, the "roundabout dog" painting by Swedish artist Lars Vilks has had consequences on the Internet. Swedish websites were allegedly attacked by hackers from Turkey and other countries. Computer teams from Sweden counterattacked.

A couple of weeks before the terrorist attacks on September 11, 2001 the pro-Taliban website *taleban.com* was hacked by a western activist.[250] The reason was that the Taliban had announced they were going to shut down the Internet in Afghanistan and all users would be punished. The activist wrote various obscenities on the website directed against the Taliban and al-Qaeda leader Osama bin Laden. Even the website *afghan-ie.com* was attacked and forced to shut down.

---

249 Cochran, A. (2006) "Muslim Hackers Assaulting Websites since Cartoon Controversy Began". Counterterrorism Blog, February 2006. On www.H-Zone.org information are published on the defacement attacks on the Danish websites.
250 "Afghan Taliban website hacked as Internet outlawed". 27 August 2001. Internet article.

September 11 created a strong feeling of indignation among many Western hacker groups.[251] Six days after the attack the German *Chaos Computer Club* (CCC) issued an appeal on their website urging hackers to avenge the attacks through DDoS attacks against targets in Afghanistan and Pakistan. They also encouraged people to steal sensitive data from various systems. CCC is one of the oldest and biggest hacker groups in the world. One of the founders of Wikileaks, for instance, allegedly has a background in Chaos Computer Club.

A short time after this, the European group *Yihat* (Young Intelligent Hackers Against Terror) announced it had hacked computers belonging to Al Shamal Islamic Bank in Sudan, including bank accounts that allegedly belonged to al-Qaeda. The bank is a part of Saudi conglomerate Dar-al-Maal al-Islami (DMI) Trust and led by Prince Mohammed al-Faisal al-Saud, a cousin of King Fahd. According to the U.S. State Department, one of the majority owners of the bank was Osama bin Laden, who used Al Shamal to transfer funds to followers all over the world.[252]

If Yihat really did manage to hack into bank accounts is not clear. Yihat's Internet manifesto describes the goal of their activity as finding terrorists, identifying them, hacking their systems and providing information to law enforcement agencies.[253] According to Yihat's own information the number of members in 2001 was about 800. That is a pretty high figure, the plausibility of which can be discussed.

Yihat's way of designating potential and possible terrorist on the Internet has caused indignation in parts of the hacker community. Both individuals and groups say they have been wrongly identified as cyber antagonists; especially Pakistani hackers have counterattacked.[254] One website under Yihat's domains, www.kill.net, was attacked by Pakistani hacker Fluffi Bunni and forced to shut down, which has resulted in Yihat becoming more careful and secretive on the Internet. Fluffi Bunni is seen as a prominent figure among Pakistani hackers, along with Dr Nuker. Dr Nuker has headed one of the most prominent groups in the country. Many advanced and serious attacks, especially in the beginning of the 21st century, are said to have his distinguishing-marks.

---

251 Marturion, D. "Will hackers keep the cyber peace?". *New Business News*, Internet article.
252 Napoleoni, L. (2004).
253 Http://www.uk/cip/resources/nipc/cyberprotestupdate.
254 Http://attribute.org/news/content/01-10-22001.html.

Chauvinism

Several of these incidents have nationalist chauvinist overtones. One of the most serious and dangerous conflicts on the Internet is the one between Pakistani and Indian hacker groups, which has been fought between different groups since 1998.[255] The Indian nuclear test can be described as the starting point.

Shortly after India's official announcement of the testing of Pokhran II, the *Milworm* hacker group attacked the Bhabha Atomic Research Centre's web page. In a defacement attack they changed the information on the site into an anti-Indian message. The group also succeeded in hijacking e-mails on the research centre's network. The identity of the hacker group is somewhat unclear. Some people say the members are based in the U.S. and Australia, others that they come from Pakistan.[256]

A little later, another group of Pakistani hackers launched an operation against the Indian Army's web site. This time they used *social engineering*.[257] Those in charge of the web site were contacted by telephone and asked to move it to another IP address. When this was done, the hacker group changed the contents of the site to anti-Indian slogans. One reason they succeed was because the server that controlled the site was not under Indian control, it was located abroad. This case shows the danger in not having control of servers with sensitive information.

The number of Pakistani defacement attacks against Indian web sites increased from four in 1999 to more than 150 in the spring of 2001, according to the Indian Computer Emergency Response Team (CERT India).[258] The number of Indian attacks against its adversaries was seven in 2000 and grew to 18 in 2001. In 2002–2004 the situation stabilised and the conflict petered out.

The first "real cyber war" between hackers from the two countries ended in a cease-fire when members of the Pakistani groups *Pakistan Hackers Club* (PHC)[259] and *G-Force*, and the Indian *NEO* agreed to discon-

---

255 Jamshed, A. (2008) "Cyberwars between India and Pakistan".
256 Chowdhuri, Satybrata Rai. (2002) "Subversive activities through cyber space". *The Tribune*, 20 May 2002.
257 In this context the term social engineering means that persons through telephone conversations acquire sensitive information such as passwords to be used to get into others systems.
258 CERT India.
259 PHC – Pakistan Hackers Club.

tinue the fighting. During the five-year conflict, hundreds of sites on both sides were attacked. More than 150 hacker groups and hackers were involved on the Pakistani side and at least 10 on the Indian.[260]

In November 2008 a new cyber conflict broke out. The reason was the terrorist attacks in Bombay, which were initiated from Pakistani soil. In connection with this, the Indian group *HMG* attacked the web site of the Ministry of Oil and Gas in Pakistan (OGRAS), which went down for several minutes. In reply, one of the now leading hacker groups in Pakistan, the *Pakistan Cyber Army* (PCA),[261] defaced at least five Indian web sites, among them Indian Oil and Natural Gas. The PCA also issued a warning on the Internet to HMG and the Indian hacker community to stop their activities.

In just a few days a large number of sites were hacked on both sides. According to themselves, the Pakistani group *KSA* managed to penetrate several of the biggest banks in India, such as Bank of Baroda. The Indians threatened to escalate the conflict by attacking the Pakistani Internet system. The conflict continued for a couple of weeks and ended in late November. An agreement was made between Pakistani groups PCA and KSA on the one hand, and *Indian Cyber Warriors* (ICW) and HMG on the other.[262] In December 2008, there was a minor incident when the Pakistanis accused the Indians of having tried to compromise a forum discussing the Pakistani armed forces.[263] The incident caused an automatic chain reaction of accusations from both sides that the cease-fire had been broken.

It is important to stress that the defacement attacks were by no means sanctioned by the authorities in either Pakistan or India, rather the opposite. Both countries do their best to avoid and counteract hacktivism at government level. Organisations in charge of network security – such as the National Informatics Centre, NIC,[264] in India and the Federal Investigation Authority, FIA,[265] in Pakistan – work actively to counteract an escalation and reduce the effects of cyber antagonism.

---

260 Raza, M-A. (2008) "India-Pakistan in State of Cyber War". 24 November 2008. Internet article.
261 PCA – Pakistan Cyber Army.
262 ICW – Indian Cyber Warriors/Hindu Militant Group.
263 Http://www.pakistanidefenceforum.com.
264 NICS – National Informatics Centre.
265 FIA – Federal Investigation Authority.

Chinese nationalism

There are number of nationalist hacker groups in mainland China. Their ambition is to support the interests of the regime and defend the honour of the country. Any connections or relations to the Party are hard to prove. The number of organised patriotic hackers is approximately 20,000. Apart from these groups there are individuals in China and abroad that act more or less independently. Many of the hackers are believed to be students and teachers at universities, as well as people with good Internet skills working in the IT business. There are probably some contacts with civilian and military research centres under the command of the Ministry of Public Security, MPS.[266]

Hacktivists operate anonymously on the Internet, they co-ordinate their activities through websites and select targets to attack. The groups are assembled and act ad hoc; they are set up when needed and disbanded when the operation is completed. Some groups are more well-organised and better structured than others. Targets may include companies, individuals and agencies in a specific country. The Chinese authorities tolerate the phenomenon as long as China or Chinese interests are not attacked.

Examples of nationalist groups include the *China Eagle Union* (CEU), *Red Hacker Alliance* (RHA), *Javaphile, China Youth Hacker Alliance, Honker Union* and *Titan Rain Hackers*. They range from grey to black hat hackers.[267] Javaphile, for instance, is believed to consist of students at Jiao Tong University in Shanghai.[268]

Red Hacker Alliance is interesting since it differs from the others; it is an official network security organisation consisting of patriotic Chinese with very good information technology skills.[269] It has a paid staff, with security experts from the universities. RHA is said to have access to recruiting officers whose task it is to identify patriotic hackers in the provinces and cities. The group has attracted hackers that have not previously been organised.[270]

---

266 "Some Experts Question Activities of China's Patriotic Hackers". *World News Connection*, 10 July 2007.
267 There are different categories of hackers; white hat hackers acting legal and testing system security. They often publish their results in public. Grey and black hat hackers acts more criminal on the net in order to steal information, to penetrate system etc.
268 The Dark Visitor. 4 April 2008. Internet article.
269 "China's Restless Hackers". *Strategypage, Information Warfare*. Article Index, 18 April 2006.
270 "Chinese patriotic hackers billed as network security research organisation". *East-Asia-Intel Reports*, 4 December 2006.

Chinese hacker *Xiao Chen* in an interview with CNN said there are several websites hosting more than 10,000 registered users. The number of nationalist websites and forums discussing hacking is more than 250.[271] There are also several hacker schools on the Internet. Tuition fees vary. One hacker called *Dark Angel* or *Hei Haitang* offers 181 individual courses at a cost of 17 dollars each. The biggest school for online hackers is reportedly *3800.com*, founded in 2003. It has a head office with a staff of 21, nine part time employees and 17 technicians. 3800.com has ten servers scattered over different regions such as Hangzhou, Yangzhou, Guangzhou, Henan, Beijing and Shanghai. Apart from these, there are probably also a number of Internet communities that teach hostile activities on the Internet.

Whether freelance hackers get financial support from Beijing or not is a matter of dispute.[272] In an interview on CNN, one hacktivist mentioned they got paid, a statement that was later denied.

A test of the level of nationalism is the anti-CNN forum that was started in 2008. The aim of the website is allegedly to retort what is seen as lies from the West. The number of individual users is said to be 500,000 per day.

A cyber conflict may be spontaneous, often initiated in connection with political events. The first major incident between the U.S. and China was caused by the unintentional American missile attack against the Chinese embassy in Belgrade in 1999, during the U.S. intervention in the aftermath of the civil war in Yugoslavia. As a reaction to what was seen as an assault on Chinese values and pride, a series of cyber attacks were launched against websites of American authorities, among them the U.S. embassy in Beijing. American hacktivists launched counter-attacks and hacked the corresponding Chinese websites. During the time that the battle lasted, more than 300 websites were attacked on both sides. What side won is unclear, some say China, others the U.S. Irrespective of this, the incident is a reminder of the danger of hacktivist actions. It should be mentioned that Red Hacker Alliance (RHA) was formed less than twelve hours after the American missiles hit the Chinese embassy in Belgrade.

Another incident occurred in 2001, when a Chinese military aircraft deliberately collided with an American signals intelligence aircraft, which was forced to land on Chinese soil. The event created ripples on the Internet.

---

271 Vause, J. (2008) "No site is safe". *CNN.com*, 7 March 2008.
272 Vause, J. (2008).

According to one source, more than 80,000 patriotic Chinese hackers were involved in attacking American interests and defending their own.[273]

Nationalist cyber attacks from China occur on a regular basis, often as a reflection of the current security policy situation. A number of controversies have been played out on the Internet between hacktivists from China and their counterparts in countries such as Taiwan and Japan. The disputes often have historical background, in some cases they are caused by territorial disputes, such as over the Senkaku, or Diaoyu, Islands in the East China Sea. Real or perceived historical injustices from before and during the Second World War also play a great role. Hackers from the U.S. and China keep testing one another's capacities to attack and defend. One incident that led to a lot of attention was the Google affair, sometimes called operation Aurora, in the winter of 2009–2010.

At the end of 2009, Google discovered several hacking attempts of their password system. The attacks were concentrated to e-mail accounts of Chinese dissidents. Apart from Google, more than twenty American Internet companies were subject to espionage in connection with this. Forensic studies indicate that those behind the operation were a Chinese top university and a vocational school for computer engineers linked to the Chinese armed forces.[274] According to Sean Noonan, analyst at American thinktank Stratfor, the reason for the operation was that Li Changchun, the fifth most powerful person in the Chinese communist party and head of the ministry of propaganda, was incensed over not very flattering information about him on Google. He reportedly ordered attacks on the search engine's Chinese web site. There is some uncertainty as to the accuracy of this information.[275]

Noonan stresses that Chinese nationalist hacktivism is a double-edged sword for the regime. While these kinds of hacker groups may benefit the purposes of the regime, there is a risk of the opposite; hacktivists turning against their leaders in Beijing.[276]

---

273   Carr, J. (2010).
274   "spionage, 1, 2010". Säkerhetspolisen, 17 May 2010.
275   Noonan, S. (2010).
276   Noonan, S. (2010).

Russian hacktivism

There are suspicions that nationalist groups with computer-skilled individuals from Russia are conducting malicious cyber activities, themselves or by proxy. One group allegedly involved in incidents on the Internet is the *Nashi Youth Group* (Democratic Anti-Fascist Movement "Ours!").

The 120,000-member strong organisation was officially formed by Vasilii Yakemenko on 1 March 2005.[277] The group was ostensibly formed to stamp out Nazi elements. There is reason to assume that it receives direct subsidies from the Kremlin and through its contacts with first deputy chief of the presidential staff in the Kremlin, Vladislav Surkov. He has met the movement several times, giving lectures and holding private talks with leaders of the movement.[278] Businessmen looking to ingratiate themselves with the regime are said to fund the youth movement.

The organisation has been accused of acting aggressively, even brutally, against opponents of the regime. The accusations include harassment, spying and physical violence; the organisation is active both in the real and the digital world. In April and May 2007, Nashi members protested daily outside the Estonian embassy in Moscow against the move of a statue of a Soviet soldier in Tallinn to a military cemetery nearby. During this period there were large-scale cyber attacks against websites and servers in Estonia.

In an interview in the *Financial Times*, Nashi activist Konstantin Goloskolov confirmed that the group was behind the cyber attack against Estonia in the spring of 2007.[279] Whether this is true or not is much debated.

In October 2007 another youth group, the Eurasian Youth Union (ESM), which was formed as a reaction to the Orange Revolution in Ukraine, conducted a distributed denial of service attack against the Ukrainian president's website, which was shut down for three days.[280]

---

277 Wapedia. http://wapedia.mobi/nashi_(youth_movement).
278 Young, C. (2007) "Putin's Young Brown Shirts". *The Boston Globe*, 10 August 2007.
279 Shactman, N. (2009) "Kremlin Kids: We Launched the Estonian Cyber War". *Wired*, 11 March 2009.
280 Carr, J. (2010).

Hacktivism changes

The development of global hacktivism is fuelled by political, religious and national manifestations and tends to increase over time. The level and intensity of the activities are governed by current events, in which conflicts spread from the real to the digital world. Symbols are often used to trigger and escalate conflicts. So far, most attacks between cyber groups have been relatively harmless, with defacement of the opponent's web sites, manipulation of contents or DDoS attacks that shut them down. Naturally, these kinds of activities are annoying for those who are subjected to them, often third parties that have nothing to do with the conflict.

However, there is a concern that the opponents will no longer content themselves with defacement attacks, raising the stakes but also the risks. There are fears that activists will conduct large-scale and co-ordinated cyber operations against sensitive targets. This is especially serious for societies that depend on working IT infrastructure. For instance, in connection with the terrorist attacks in Bombay, Indian hacker groups threatened to launch attacks against vital IT infrastructure in Pakistan. This development could turn ugly in a short time, resulting in a situation that is hard to control. There might be a more or less unintentional escalation.

One hacker group that has gained a lot of attention is *Anonymous*. The group has become something of a symbol of hacktivism and has been active in several arenas on the Internet. For instance, in connection with the Wikileaks affair in the autumn of 2010, where Sweden wanted Julian Assange deported from Britain, the Swedish office of the public prosecutor was attacked by cyber activists associated with the group. More than 500 computers were involved in sinking its website. This was a comparatively small attack. Anonymous has also defaced the FBI's website and threatened to launch attacks on a massive scale against its adversaries. In 2011, Anonymous joined the group *TeamPoison* in operation Anti-Sec, where various law enforcement agencies, security organisations and government offices worldwide were attacked. Anti-Sec also succeeded in hacking the Twitter account of Fox News. TeamPoison has also been given credit for an attack on NATO's website, defacing many of its web pages.

In the cyber security community there is a fear that Anonymous, or similar kinds of skilled hacker constellations driven by ideological causes, join forces with rogue states with a lot of resources. If this happens the logic of hacktivism could change.

A massive operation with participation by many hacktivists could lead to ripple effects with security policy consequences, and spread across borders, fast. Companies and organisations that have nothing to do with the conflict could be targeted, which the Google incident in the winter of 2009-2010 indicates. At the same time it would be very difficult to know who or what groups initiated the attack, and for what purpose. It could be part of a deception campaign or a bigger military operation. One problem, from a security policy point of view, when trying to identify the instigator, is the opportunity to conceal activities on the Internet. There is an obvious risk that wrong people are accused, and that the response is disproportionate.

The emerging cyber threats from malicious hacktivism and different kinds of hostile activities show the importance of improving both information security and international co-operation in order to prevent or at least reduce the negative effects of antagonistic cyber operations. The issue of cyber threats must be resolved on a global scale, with all major parties and the law enforcement agencies of all nations. Conventions have to be rewritten since cyber antagonism confounds principles such as proportionality, neutrality and distinction. By co-operating to make cyberspace more secure against criminal intrusion, the work would also lead to improved security for military information operations. The conclusion is that the cyber rules of engagement need to be discussed further. But in order to design viable forms of guidelines on how to behave on the Internet, it is important to understand the modus operandi of different players, the culture and logic behind their actions and their motives and driving forces.

One way of developing new knowledge of hacktivism and how it has developed is to study the cyber operations against Estonia in 2007 and Georgia in 2008. They are described in next chapter.

Towards a new modus operandi? The Estonian and Georgian cyber war experiences

The Estonian cyber conflict in the spring of 2007 has attracted a great deal of interest. Some describe it as the first official and publicly described cyber war against a country. Others say it was not a war but a cyber riot.[281] Irrespective of this, the incident has been a wake-up call for authorities responsible for information security in many countries. It demonstrates the built-in vulnerabilities in vital information systems and the need to change and improve security routines. The cyber attacks also show the risks of nationalist and politically motivated hacktivism, using symbols to set off conflicts and escalate them.

The Estonian conflict

This incident was provoked by the move of a Soviet military statue from the centre of Tallinn to a nearby military cemetery. Many Estonians see the war monument as a symbol of the Soviet occupation forces and the annexation of the Baltic States. The move caused a lot of anger among ethnic Russians in the country, setting off riots in the streets of the capital. In connection with this, an intense activity began on the Internet. An operation was started, the objective of which was to attack Estonian computer systems and a number of national websites.

According to Lauri Allman, Estonia's permanent undersecretary of defence, there were two phases to the attacks. The first was carried out at 1 A.M., 28 April.[282] Initially, relatively primitive and simple tools were used. Several

---

281 Brenner, B. (2007) "Black Hat 2007: Estonian attacks were a cyber riot, not warfare". *Information Security Magazine*, 3 August 2007.
282 Kash, W. (2008) "Lessons from the Cyberattacks on Estonia. Interview with Lauri Allman, Estonia's permanent undersecretary of defence". *Government Computer News*, 13 June 2008.

websites and Internet forums, most of them Russian, encouraged hacktivists to contribute. These sites also contained programs and instructions that could be downloaded and used in the attacks of Estonian websites. The targets included the Estonian Government Briefing Room, the Estonian Ministry of Defence and leading political parties in the country. The attack peaked around 3 May and slowly subsided after a period of general fatigue. The effect was not sufficient, due to the lack of a critical mass of people getting involved in the operation.

The second phase of the attacks peaked around 8 and 9 May, two of the most celebrated dates in the Russian calendar, when the country celebrates Victory Day over Nazi Germany in 1945. This time the tools were more sophisticated, mainly large botnets of compromised computers conducting DDoS attacks. The websites of the Estonian Parliament, two of the country's biggest banks, almost all of the country's government ministries and three of the six biggest news organisations were targeted.[283]

Within a few days, servers and networks were overloaded with information, which led to reduced functionality. Websites were forced to shut down. Some defacement attacks were also made during the operation. Mission-critical computers, for example the telephone exchanges, were targeted. This indicates that the people behind the attacks probably had inside information about specific, important systems to attack. The cyber attack stopped as quickly as it had started. The attackers stopped voluntarily, rather than be shut down.[284]

After the first indications that they were under attack, a team of people was quickly involved to start working on how to protect the country's Internet sovereignty. The Estonian CERT and private entities co-operated closely to solve the problem. There was an informal agreement to share information openly between the protectors and not compete on security. Co-ordination of resources was easy since Estonia is a small country with only about 1.4 million inhabitants and in CERT more or less everybody already knew one another.[285]

---

283   Traynor, I. (2007) "Russia Accused of Unleashing Cyberwar to Disable Estonia". *Guardian. co.uk*, 17 May 2007.
284   McAfee Virtual Criminology Report. "Cybercrime: The next wave", 2007.
285   Brenner (2007) "Black Hat 2007".

The first response by the Estonians was to increase the Internet throughput capacity in co-operation with other countries. This was done incrementally. They also tried to block external servers. During the operation, the Estonians identified several ping messages being sent in order to measure the country's throughput capacity. Intelligence shows that the result of the measures changed the behaviour of the attackers, who adjusted their actions in response.[286] It was a struggle between attacks and counterattacks. The attackers kept getting new information on how to attack and respond to defences.

A rough estimate is that the attacks emanated from 75 or more locations using 1 million or more computers.[287] At the height of the attacks more than 20,000 networks of compromised computers were part of it.[288] Analysis of the IP addresses of the attacking computers shows a long list of states from all around the world – as many as 178 different countries.[289]

It is important to point out that it is possible to fake IP addresses. Many analysts believe that the attacks were carried out by a well-organised group with good financial and intellectual resources. The attacks came in waves. At its peak, the attack measured about 100 MB of traffic per second, which is considered to be quite moderate. In comparison, the largest DDoS attacks have measured up to 40 GB per second.[290] Several Internet security experts, such as Russian Internet pioneer Anton Nossik, say that "compared to the scale of the problem in general, Estonia is small".[291]

Mike Witt, deputy director at the US CERT, also believes that, while the "size of the cyber attacks was certainly significant to the Estonian Government, from a technical standpoint is not something we would consider significant in scale".[292]

---

286 Kash, W. (2008) "Lessons from the Cyberattacks".
287 Kash, W. (2008).
288 McAfee (2007) "Cybercrime: The next wave".
289 A comment made by Katrin Pargmae, a spokesperson for the Estonian information centre, to the Homeland Security News Wire (HSNW). "2007 Cyber Attack on Estonia Launched by Kremlin-backed Youth Group". *Homeland Security Newswire*, 13 March 2009.
290 A comment made by Jose Nazario from the information security company Arbor Networks referenced in the Homeland Security News Wire (HSNW).
291 Anton Nossiks: "Compared to the scale of the problem in general, Estonia is small." Published in BBC News, "The Cyber Raiders Hitting Estonia", 17 May 2007.
292 Mike Witts: "The size of the cyber attack, while it was certainly significant to the Estonian government, from a technical standpoint is not something we would consider significant in scale." Published in United Press International (2007) "Analysis: Who Cyber Smacked Estonia?"

In general, for moderately computer-skilled persons it is not too difficult to lease botnets with a large number of compromised computers for malicious activities such as DDoS attacks. The rental cost for botnets is somewhere between 1,000 and 5,000 U.S. dollars.[293] One interpretation as to why the DDoS attacks stopped was that the rental time of the leased botnets was over. By that time the instigators of the attacks had achieved their goal. The message delivered to the Estonians was clear.

It has not yet been fully established who or what groups and organisations were behind the operation. Some of the IP addresses indicating servers are probably fake. There is no indication of Russian government involvement. Persons linked to Nashi have said that they were behind the operation, but the statement is disputed. State Duma member Sergei Markov claimed that one of his assistants was responsible for instigating the cyber attack in Estonia.294 This should be interpreted as a provocation; it is not necessarily true.

Estonia is one of the world's most connected countries, and therefore more vulnerable to cyber attacks than less modern societies. For instance, more than 97% of all banking transactions are made online. The cyber attacks show the built-in vulnerabilities and the need for investments in cyber security.

One effect of the attacks is that Estonia has reinforced its cyber emergency response team. In co-operation with NATO, a cyber security centre was set up in August 2008 (the official name is Co-operative Cyber Defence Centre of Excellence – CCD COE, usually referred to by the codename K5). In total, a group of 30 experts are permanently stationed in the Tallinn area.[295] Estonia is working on a national cyber defence strategy. It involves factors such as making the backbone Internet infrastructure more robust and expanding the Internet throughput capacity.[296] Other areas are investments in capacity to detect cyber attacks. A method to monitor and control databases used for e-services by the government has been implemented at central level, called *X-Road*.

---

293 Francis, B. (2005) "Hacker Sells Their Information Anonymously through Secretive Websites. Know thy hacker". *Infoworld*, 28 January 2005.
294 "Transmission. Behind The Estonia Cyberattacks". 8 March 2009. Internet article.
295 Johnson, B. (2009) "No One Is Ready for This". *The Guardian*, 16 April 2009.
296 Kash, W. (2008) "Lessons from the Cyberattacks".

The cyber operation against Georgia

The cyber attack against Georgia in the summer of 2008 has, together with the attack on Estonia, been a wake-up call highlighting the risks, threats and vulnerabilities of information warfare. New insights have been gained about which means and methods an aggressor can use to conduct computer and network operations in combination with psychological and military operations against an adversary. Basically this was the first time that an online operation was combined with a military offensive.[297]

The event shows some interesting features both when it comes to the way in which it was prepared and conducted, and what consequences it had. A new modus operandi can be discerned, setting the standard for future malicious activities in cyber space. Some remarks about or characteristics of the operation are the following.

In June 2008, almost two months before the actual start of the five-day military conflict between Russia and Georgia, the first small-scale DDoS attacks occurred.[298] They were carried out by botnets using zombie computers infected with malware, specifically constructed to attack designated targets.

On 20 July, multiple DDoS attacks were registered by the *Shadowserver Foundation*, an Internet watchdog group of volunteers specialising in malicious online activities. The attack was aimed at the official website of the Georgian president, Mikheil Saakashvili, which was forced to shut down for 24 hours. Analysis showed that the attack was directed by a command and control server based in the U.S. The server was set up just weeks before the actual conflict started.[299]

On 8 August, the same day as the military offensive started, with Russian forces moving across the Georgian border, the websites of the president of Georgia, the Georgian Parliament, the ministries of defence and foreign affairs, the National Bank of Georgia and the online news agencies were at-

---

297 "Recent Events Suggest Cyber Warfare Can Become New Threat". *WMD Insights*, December 2008/January 2009 issue.
298 Hart, K. (2008) "Longtime Battle Lines Are Recast in Russia and Georgia's Cyberwar". *Washington Post*, 14 August 2008.
299 Markoff, J. (2008) "Before the Gunfire, Cyberattacks". *The International Herald Tribune*, 13 August 2008.

tacked by hacktivists. The websites were forced to shut down. Shadowserver detected that the first co-ordinated online assault was run by several different botnets. The number of cyber attacks escalated as the military conflict became more intense.[300]

The military conflict between Georgia and Russia has several dimensions and goes back to a number of events, such as the Georgian application for membership of NATO. There were provocations on both sides. The war started when Georgian troops surrounded Tskhinvali, the capital of South Ossetia, which had requested independence from Georgia, with the support of Russia. Many inhabitants in South Ossetia had Russian citizenship, Russian troops invaded Georgia.

Website defacements were also conducted as part of the psychological pressure aimed at discrediting Georgian President Saakashvili. Images on the president's personal website were digitally manipulated to make him look like Nazi leader Adolf Hitler.[301] The defacement of the website was very skilful.

The Georgian government tried to counter the aggression in different ways. Installing attack filters to block Russian IP addresses was one method used to reduce the effects of the DDoS. Another was to move the websites of the Ministry of Foreign Affairs and civil.ge to a blogspot domain that was better protected.

As the cyber conflict escalated, contacts were made with Estonia and other countries and organisations to help reduce the effects. Estonia dispatched information security specialists from the national CERT to help Georgia defend its cyber sovereignty. Poland also helped the Georgian government by providing Polish websites that could be used by the Georgian authorities to dispatch information on their view of the hostilities.

Project Grey Goose 2, an open source intelligence (OSINT) initiative led by cyber analyst Jeffrey Carr, has tried to answer the question of whether the Russian government, or groups loosely linked to it, was involved in the cyber operation or if it was the work of a grassroots hacker movement alone.[302]

---

300 Waterman, S. (2008) "Analysis: Russia-Georgia Cyberwar Doubted". *United Press International*, 14 August 2008.
301 Porthillo-Shrimpton, T. (2008) "Battle for South Ossetia Fought in Cyberspace". *The Independent*, 17 August 2008.
302 Project Grey Goose Phase II Report. "The evolving state of cyber warfare". 19 March 2009. Internet article.

The method the project used to get an idea of possible links was based on semantic analyses of hacker blogs discussing the Georgian issue. By searching Internet forums and blogs, the Grey Goose team members could collect information on the "kill chain" – how novice hackers were recruited to participate, the development of target lists, the selection of malware to be used and finally the decision on how to launch the attack.[303]

Other organisations, such as the US Cyber Consequence Unit (US-CCU), have also studied the issue. The result of their investigation is confidential.[304]

The Grey Goose project identified two Russian hacker forums as instigators, where the attacks were organised during the operation – *stopgeorgia.ru* and *Xakep.ru*. For instance, stopgeorgia.ru was set up within hours of the Russian armed forces invading South Ossetia. Information was constantly updated in the forum in order to instruct potential hackers on how to attack Georgian sites. Lists of target websites were featured and visitors were encouraged to download a free software program, which allowed them to participate instantly in the massive DDoS attacks.

The stopgeorgia.ru website used an IP address connected to a hosting firm called Steadyhost (www.steadyhost.ru). Although the Steadyhost operator is registered in New York, it operates out of St Petersburg. The interesting thing about it is that Steadyhost is believed to have its offices in the same building as a Ministry of Defence institute, the Russian Centre for Research of Military Strength of Foreign Countries.[305] The GRU's headquarters is also situated on the same street.[306]

As a way of supplementing the DDoS attacks, SQL injections (junk code that confuses a website's back-end database) were used to exploit application vulnerabilities in MySQL software. Millions of junk queries were sent, overwhelming the target databases, making the corresponding servers inoperable.

Compared to DDoS attacks, SQL injections are difficult to detect and require fewer computers to achieve the same objective as DDoS attacks. The Grey Goose report establishes that the SQL injection attack "shows

---

303 Matthews, W. (2008) "New Ways of War: Cyber attacks likely in any military conflict". *Defence News Reports*, 26 October 2008.
304 US-CCU (2009) "Overview by the US-CCU of the Cyber Campaign against Georgia in August of 2008". Special report.
305 Ministry of Defense institute, the Russian Centre for Research of Military Strength of Foreign Countries.
306 Http://intellibriefs.blogspot.com/2009/03/cyber-warfare-project-grey-goose-phase.html.

moderate technical sophistications, but more importantly, it shows, planning, organisation, target reconnaissance, and evolution of attacks".[307]

Towards a new modus operandi

The operation against Estonia was one of the first official and publicly known cyber attacks against a country using large-scale botnets and DDoS by nationalist civilians. In the Georgian operation, the methods were even more refined.

Both cyber operations were well co-ordinated in time and space and the attackers seemed to know what kind of websites to attack and how to take them down. This implies that computer network exploitation and other reconnaissance methods were used in advance. In the Georgian case the actual cyber operation was initiated and conducted in conjunction with the military offensive.

The Georgian operation was carried out by civilians – nationalist-oriented individuals and groups of people, possibly with the support of cyber criminals such as the Russian Business Network, RBN. Social networks were the main tool for recruiting potential hacktivists and for providing malware to the hackers. Basically three methods were used by the attackers – distributed denial of service attacks, SQL injections and website defacements. These were relatively unsophisticated kinds of attacks, but they were carried out in an innovative way.

The targets were government and news media websites as well as Georgian financial and educational institutions. The cyber attacks reduced the ability of the Georgian government to counter the Russian invasion. The defenders' resources had to be split between different activities and areas. Besides that, the operation had psychological impact, in the sense that it interfered with the government's ability to communicate with the public. The co-ordination between the cyber campaign and the military offensive is probably not a coincidence. Any connection to the Russian authorities is very hard to prove, in both cases. The Russian government denies any accusations of participation, and there is no evidence that it initiated or conducted the campaigns.

---

[307] The quote in the report is the following: "... the SQL injection attack ... shows moderate technical sophistication, but more importantly, it shows, planning, organization, targeted reconnaissance, and evolution of attacks".

The events can be seen as a new modus operandi that may set the standard for future cyber conflicts. In theory, it would be possible for a player to use nationalist hackers, thus gaining deniability and the ability to enjoy the strategic benefits of their actions, but not sharing the risks. Moreover the cyber weapon can be used in order to put psychological pressure on opponents to act in a way that is detrimental to him.

The impact of the cyber operation against the Georgian communications and information infrastructure was limited due to the low Internet penetration in the country. But it demonstrates the possible effects that could be achieved. The consequences of a well-co-ordinated cyber operation against critical systems and networks in more advanced countries that depend on modern information infrastructure would be far more serious, as the Estonian case shows. The implication of the new modus operandi stresses the need for improved information security at all levels of society, as well as the need to co-operate at international level to reduce the tensions and effects of cyber attacks. In next chapter the need for regulation of malicious activities in cyberspace is discussed.

The need for a treaty in cyberspace

Comparing cyber weapons and weapons of mass destruction, as some Russian analysts have done, is quite drastic. If a country that believes itself to be under the threat of a cyber attack were to resort to a deterrence strategy, this would lead to a dangerous escalation of the situation, especially bearing in mind that it is very easy to conceal digital traces and mislead adversaries in cyberspace.

Such a development could, in a short period of time, lead to an unpleasant situation that would be hard to control or manage, with more or less unintentional escalations. In view of the cyber attacks against Estonia in 2007 and Georgia in 2008, there is a growing concern that different kinds of activists could and would carry out large-scale and co-ordinated cyber operations against vital targets.

A number of questions have to be answered: Is the response to a potential cyber operation a task for law enforcement agencies, or a matter for the military or some other organisation?[308] Should it be carried out and resolved on national level or internationally? And with what means? According to UN Article 51 every nation has the right to protect itself against armed attacks. Is this the case in cyberspace, too?

How should phenomena such as cyber terrorism, cyber crime and cyber espionage be handled? One tricky issue is how to deal with non-state hackers involved in every aspect of cyber aggression while providing plausible deniability to the host governments. For instance, the activities of "black hat" hackers are not limited to a specific area; they cover a wide range, the whole scale of malicious activities, from cyber crime to cyber warfare. Moreover, third parties can be used both as an aggressor as well as for deception.

---

308 "The Russian view is that the USA see international information security as a way to control criminality in the information domain and chose to forget the existence of information warfare weapons" according to Komov, S.A., Korotkov, S.V., Rodionov, S.N. (2003) "International Information Security: Military aspects". *Military Thought*, vol. 12, No. 4, 2003. Komov, S.A et al. (2003).

These questions stress the need for a "common code of conduct" and agreements between all major nations on how to behave in cyberspace, and the level of response if an attack occurs. There is a need for regulations and operating procedures that provide guidance on how to act in order to limit the consequences. The problem is what should be regulated, how it should be regulated and in what form this should be done.

Different views

Cyber aggression has legal and political aspects. There is a fundamental difference between the Russian and U.S. views on the need to regulate hostile activities on the Internet. The U.S. view is that a treaty is unnecessary. Instead the U.S. advocates improved co-operation among international law enforcement agencies such as Interpol and Europol.

By co-operating to make cyberspace more secure against criminal activities, their work will also lead to improved security for military elements.[309] The U.S. also rejects any agreements that would allow governments to censor the Internet, something that would favour totalitarian regimes. The distrust is primarily aimed at Russia and China and their view on control and supervision of their citizens.

However, the U.S. view on cyber regulation is changing. A new strategy was introduced in May 2011, which means that the U.S. reserves the right to counterattack digital assaults against infrastructure that is vital to society. This changed attitude will affect what is regarded as a criminal act and what is military aggression. Furthermore, the U.S. has encouraged NATO to stop cyber attacks. A major attack against one member of the alliance should be seen as an attack on all of them.[310]

The Russian view is the opposite. From a Russian perspective, the absence of a treaty permits a kind of arms race in cyber space that may have unforeseen consequences. From the Russian perspective, the IW weapon should be included in disarmament negotiations and handled like treaty limits on troops (forces, numbers, equipment etc.). Russia has proposed a disarmament treaty that bans countries from secretly developing malicious code and

---

309 Markoff, J., Kramer, A. (2009) "U.S. and Russia Differ on Treaty for Cyberspace". *The New York Times*, 27 June 2009.
310 Cyber 3.0. U.S. Department of Defence (2011).

built-in defects in computers, which can be activated in the event of war.[311] Other Russian proposals are linked to agreements and conventions on human rights, banning attacks on non-combatants and deception in operations in cyberspace. The latter is an attempt to manage anonymous attacks. An inevitable observation is that a ban on the development of malicious code will be very hard to supervise.

Russia has been active in this field for several years. In 1998, UN General Assembly Resolution No. 53/70 was drafted on a Russian initiative.[312] In 2009, a Group of Governmental Experts (GGE) was set up by UNIDIR[313] to look into the impact of information and communications technology on international security.[314] The framework of the Yekaterinburg convention from 2011 between Russia, China, Tajikistan and Uzbekistan is also an active step to bring the Russian view to the international community. The reasons for the Russians' involvement in the work of UNIDIR and other forums are complex.

One reason is that they really are concerned about the effects of massive cyber attacks on critical information infrastructure. An operation could disrupt the state administration system, demoralise the population and destroy or disable key elements of the important military-industrial complex. The psychological consequences, and the economic and military implications, could be severe. Dangerous situations could arise in a very short time, spreading to other areas, and would cause pressure at security policy level. The Russians remember the disintegration of the Soviet Union.

Another interpretation of the Russian involvement is that both the U.S. and China, Russia's primary competitors, are investing heavily in information warfare capacities, e.g. the development of an American Cyber Command under the Strategic Command and the build-up of the Chinese Information Operation Corps and Cyber Militias. Bearing in mind that both the U.S. and China are major suppliers of hardware and software worldwide, the Russians fear back-door functions and logic bombs hidden inside computers and networks. Russia, on the other hand, does have very skilled programmers and competitive software companies in the information security area, many of them work for American IT companies.

---

311 Markoff, J., Kramer, A. (2009) "U.S. and Russia Differ".
312 Komov et al. (2003) "International Information Security".
313 The United Nations Institute for Disarmament Research.
314 ICT and International Security (2007) Disarmament Forum, 3. UNIDIR.

The Russians may have the feeling that they are behind their competitors. In their experience, regulation could be one way to gain control over the progress their opponents are making. S.A. Bogdanov e.g. says that during the last few decades, Russia's military-economic capacity has seriously weakened and the army "only has 20 percent modern weapons and military equipment".[315] Additionally, "the Russian military will have to fight with weapons that are qualitatively inferior to those of possible adversaries on a number of parameters, especially regarding communication and intelligence systems as well as electronic warfare (EW) and precisions guided weapons". The statement could of course be a part of a *maskirovka* in the sense that the Russians are presenting themselves as weaker than they actually are in order to win time to build up their resources. Information from the Georgian five-day war shows that Russia to some extent lacks sufficient EW equipment.

A third reason could be that the Russians want to act proactively in order to tone down the on-going discussions in various forums indicating that activists from Russia were behind the cyber attacks against Estonia in 2007 and Georgia in 2008 – suggestions which Russia strongly denies. There is no evidence that Russian authorities or groups linked to them were involved in the cyber conflicts, although unconfirmed information on the Internet indicates that people linked to the Russian authorities may have encouraged patriotic hackers to act in the early phases of the Estonian and the Georgian conflicts. By getting involved in regulatory bodies for cyber protection, Russia could win political points. Major-General Alexander Burutin, acting deputy chief of the General Staff, has mentioned the need to establish an "Agency for a positive image of Russia" to check negative attitudes to the country.[316]

This said, however, there are some areas where Russia is not keen on regulation. For instance, a proposal to regulate cyber crime under a UN directive is still under consideration by Russian authorities. One reason for the delay could be that many of the criminal activities conducted on a large scale worldwide originate from Russia or are directly or indirectly linked to the country. The infamous Russian Business Network, RBN, is said to be

---

315 Bogdanov (2004) "Warfare of the Future".
316 Project Grey Goose (2008) "Some lessons learned, and more to come". *IntelliBriefs*, 22 October 2008. Internet article.

the mother of all cyber crime.[317] There is a suspicion that there are some links between RBN, Russian authorities and organised crime.[318] On the other hand, there is information indicating that RBN has been dissolved.

China's official view is unclear when it comes to a treaty – what kind and what scope? What is happening is hard to interpret. The regime's overall information security policy is to encourage "stability" and "harmony" at all levels of society. This can be achieved through censorship, surveillance and control of citizens' activities on the Internet. Officially, the regime refers to the Yekaterinburg convention. A term that is used off and on is "Freedom of Information" – to be understood as the regime's focus to create an internal environment with "secure" information. The opinion is that every country is responsible for its own information protection; nobody should tell China how to solve its internal affairs.

The regime has had a tolerant attitude to nationalist hacker groups, as long as they are not acting against national interests. But this view is beginning to change, at least partly, due to the insight that hackers can assume different roles and positions depending on the situation. A clear trend in China is increased levels of cyber crime aimed both at Chinese and foreign interests. The regime has realised that the cyber weapon is a two-edged sword that can also be directed against China. Hackers have been brought to justice. So the issue is getting more and more attention. Whether China wants to control harmful cyber activities, including espionage over Internet and different aspects of cyber wars, and if so in what way and in what forums, is less clear.

Whatever the explanations or reasons for getting involved, or not, in organisations such as UNIDIR for cyber arms control, cyber crime, cyber terrorism and cyber espionage, it is very important to create some sort of international co-operation in order to prevent or reduce the negative effects of antagonistic cyber operations. This issue has to be solved on a wide front, involving all major parties, nations and law enforcement agencies.

Conventions have to be rewritten since cyber war confounds traditional principles such as proportionality, neutrality and distinction.[319] "Cyber

---

317 O'Connell, K. (2007) "INTERNET LAW: Russian company outed as mother of all Cybercrime". 24 October 2007. Internet article.
318 Flook, K. (2009) "Russia and the Cyberthreats". *Critical Threats*, 13 May 2009.
319 O'Connell, K. (2007).

rules of engagement" have to be discussed. One question is how to define acceptable cyber rules similar to the ones for war.

There are many areas to be addressed and resolved. An agreement on cyberspace will have to deal with issues such as censorship of the Internet, sovereignty, and how to handle rogue players who may not be subject to a treaty. It must also include all forms of network and digital activities, not just on the Internet and in cyberspace, but also covering fields such as electromagnetic pulse weapons and other related areas. Openness, transparency and freedom of speech on the Internet actually have to be noticed.

For each individual state it is important to follow the discussion in progress in different international forums and, based on that, work out its own attitude, strategy and plan to fulfil the requirements.

Future cyber threats

The increasing number of threats in cyberspace is directly related to the information technological development. Looking at the future, up until around 2030, it is not controversial to state that the information technological revolution will continue at unreduced pace and more and more people will be connected to the global network.

As populations grow, economic conditions improve and applications and technology get cheaper, the number of Internet and cell phone users increases. The development is characterised by mobility in services and access forms; users no longer depend on specific terminals, times or geographical locations to communicate. Communication takes place and information is spread wherever and whenever. Connections are ad hoc and decentralised; a connection is set up when needed and closed down when the session is over.

Every individual with access to communications equipment of some kind will also be part of a universal communications infrastructure, similar to the American military vision of the "Global Information Grid" (GIG). This also applies to devices used for machine-to-machine interaction. Telecommunications company Ericsson predicts that the number of units connected to Internet in 2020 will be 50 billion.[320] Machines communicate with other machines, automatically, without supervision by humans. This involves all kinds of electronics built into cars and homes, for surveillance and alarm, that send signals over the Internet.

The trend is towards better processors, broadband access and embedded intelligence in networks, platforms and services. Thanks to more advanced and autodidactic algorithms, the web becomes smarter. Integrated intelligence gives the system the opportunity to automatically analyse and package information, and distribute it to users depending on their situation and requirements. The vision is known as *Web 3.0* or *semantic Web*.[321]

---

320 "Towards 50 Billions devices connected?" Internet article.
321 Segaran, T., Evans, C., Taylor, J. (2009) "Programming the Semantic Web". *O'Reillys & Associates.*

The first version, called Web 1.0, was introduced in the 1990s and was about building platforms and connecting computers. The purpose was to create working links and navigation systems such as Yahoo, AOL, Netscape and Internet Explorer. The present system can be described as Web 2.0, symbolised by social networks such as Facebook, YouTube, LinkedIn and other forums and blogs.

From a risk and security perspective, the development has a number of implications. Modern society quite simply does not work without information and communications technology, where information is the most important and valuable component in the network. Just as energy; data and information is constantly changing shapes, consumed, transformed and distributed over the global network. The importance of information demands protection. Securing the contents of the information against intercepts, manipulation, deception, interruption and destruction is important to all users, irrespective of whether they are nations, companies, organisations or individuals.

According to NATO's long term study, the threats in 2030 will be complex and diverse.[322] Some threats can be extrapolated from the present situation, others will be new and unknown. The rapid development of information and communications technologies, and their fundamental consequences for human interaction, means that cyberspace increasingly will be an area of conflicts.

In the future, information warfare will be an integrated part of every major military and political conflict. Cyberspace will be both its own dimension and a co-ordinated part of the other physical arenas; land, sea, air and space. Information warfare will be conducted at all levels; strategic, operational and tactical, with different emphases during different phases. A conflict may both start and be decided in the digital sphere, without using kinetic means. Offensive and defensive information warfare tools can be used by themselves or in combination with other weapons systems.[323]

In a military context, information warfare will be a force multiplier to military operations or an individual capacity. Some military analysts define

---

322 NATO. Multiple Futures Project. Navigating towards 2030. Final report – April 2009.
323 Heickerö, R. (2010) "Cyber security challenges for Asia in a 2030 time frame". *Asian Security Conference 2030. IDSA.* New Delhi, 11–13 February 2010.

future war in 2030 as sixth generation warfare, based on intelligent networks, artificial intelligence and computers.[324]

Future information weapons include a number of technological, digital and psychological functions directed against adversaries' vital networks, command and control systems, organisations, individual operators and commanders. There is an obvious risk that the line between military and civilian structures is blurred and diluted, since all parties use joint communications infrastructure in one way or another. Attacks are direct and indirect. It will be possible to manipulate, disrupt and distort information, just as it will be possible to physically destroy everything from single components up to whole systems. Psychological operations will become an even more important feature of information operations.

For weaker parties, the cyber weapon will be attractive in contexts where they try to avoid direct military conflicts with conventional weapons. Instead, operations are conducted over the Internet. Victory over an enemy can be achieved at a low cost and without having to occupy his territory. The information weapon is also a potent instrument for deterrence. It can be used to bring psychological pressure on an opponent.

Changing roles

The development paves the way for irregular activities. There will be a number of different players; everything from individuals, insiders, financially motivated criminals, ethnically, religiously and ideologically driven groups and cyber terrorists to states with intelligence and security services.

The roles of the players will probably gradually change, since the Internet makes new patterns of interaction possible. There is a risk that unholy alliances will appear between different categories of players and antagonists. Network logic may make people who normally would not have much in common start co-operating and developing new logics and skills. There are signs that this is already happening. Social media and web forums are used as market places to spread ideologies, for recruitment and to co-ordinate operations.

---

324 There is an ongoing discussion on how to define the ongoing transformation of warfare. The sixth generation warfare is a part of an RMA including cyberwar and autonomous platforms as basic fundaments. See for instance Manwaring, M. (2006). "The Asymmetrical 21st Century Security Reality". Strategic Studies Institute. US Army War College. *IOS Press* 2006.

When it comes to cyber terrorism, it is theoretically possible that "ordinary", not radicalised, hackers are motivated to conduct aggressive actions by being offered money and prestige by terrorist groups. The probability for this scenario is relatively low.[325] The parties will have difficulties finding common areas on the Internet, their logics also differ. But the motives and driving forces may change as time passes.

"Cyber mercenaries", who charge for their services regardless of the client, are a new phenomenon. This way, warfare is privatised. From a military historic perspective this is nothing new. The difference is that in this case the fighting has moved from the physical world into the virtual one.

In the future it will be possible to imagine situations where cadres of cyber mercenaries get involved in conflicts – for money, political, national, religious and/or ideological reasons. They act as deputy players in large cyber battles.[326] Embryos of such warfare by proxy have already been seen, in the operations against Estonia in 2007 and Georgia in 2008.[327] In both cases there is no public information that the hacktivists involved were paid for their involvement.

The reason why third parties are used is so that the instigator, motives and intentions behind an operation can remain hidden. Using proxies is a good strategy for states, but also for individuals and companies that want to gain an advantage over their opponents and competitors. It is especially useful when collecting information illegally in cyber espionage.

Dissatisfied individuals and small groups of skilled hackers are assessed to become an increasing problem in future. They can be part of a criminal organisation or act politically and/or ideologically through hacktivism. In some cases they will act independently without any ties to organisations, as an army of one.

The logic of cyber crime is money. Future highly interactive networks will be a superb forum for all kinds of fraud, virtual robberies, blackmail attempts etc., directed against individuals, organisations and financial institutions. This kind of crime is part of a global movement and an international phenomenon that is not restricted by geography, national origin or laws. The cyber mafia will act like huge, resourceful conglomerates, always on

---

325   Collin, B. (1997) "The future of Cyberterrorism: Where the Physical and Virtual Worlds Converge". 11th Annual International Symposium on Criminal Justice Issues.
326   Bardin, J. (2009) "Cyber Mercenaries – Avatar Forces". 7 November 2009. Internet article.
327   Heickerö, R (2010) "Emerging Cyber Threats and Russian Views on Information Warfare and Information Operations". User Report. FOI-R—2970—SE. Mars 2010.

the hunt for new profitable markets. Infamous groups like the now disbanded Russian Business Network (RBN), which had an annual turnover of several billion U.S. dollars, shows us what to expect. These kinds of loosely compiled organisations are assessed to increase. They will also co-operate with "conventional" organised crime. Behind the crime syndicates there may be links to less scrupulous regimes.

The development of information weapons

Infological weapons, such as malicious code, are estimated to become cheaper and easier to supply to more and more people, except for the most advanced trojans and worms that take a long time to develop, which are only for resourceful players. An increased accessibility reduces the barriers for using this kind of means.

The development is heading in several directions. One direction is to use relatively simple and cheap malware in the form of "crime ware kits" against specific targets in order to destroy, compromise or assume control over the target's computers and networks.[328] They can be used for attacks against military and civilian command and control systems. Using "swarming" techniques, attack waves can come from multiple directions.[329] This kind of malware can be purchased at a low cost over the Internet.

A second direction of development is implementation of more intelligent, self-copying and autonomously driven stealth viruses that can be planted in systems and programs, and hide for years and before being activated when needed. This can be done by insiders or through users who unintentionally download malware from websites on the Internet. These kinds of intelligent and adaptable viruses have the capacity both to transform and to reproduce themselves if necessary. In order to make discovery more difficult they can also destroy themselves when the operation is over. The code is individually adapted depending on the target. The Stuxnet worm, which infected the Iranian nuclear programme, belonged to this kind of code. These kinds of operations will probably increase since they have turned out to be very efficient.[330]

---

328 Zeller, J. (2006) "Cyberthieves Silently Copy Your Passwords as You Type". *The New York Times*, 27 February 2006.
329 Edwards, J (2004) "Swarming and the Future of Warfare". Dissertation. Pardee Rand Graduate School. *Rand Corporation 2005*. Santa Monica, USA.
330 Augustsson, T. (2010) "Litet hopp att stoppa nätvirus". *Svenska Dagbladet*, 16 June 2010.

A third direction is malware with the ability to "sniff" the net autonomously searching for vulnerabilities and back door openings. They act like loose cyber dogs, either stealing, manipulating or distorting information. Just as trojans, they can also be used to take over computers and networks.

The development of machine-to-machine interaction, through which billions of units are connected to the Internet, as described in Ericsson's study of the future, constitutes great risks. In this perspective, the biggest botnets today, with a couple hundred thousand compromised computers, appear small. The botnets of the future will consist of tens of millions hijacked units.

When it comes to offensive and defensive weapons, there is concern for hostile activities in the form of "algorithm bombs" and "software bombs". This kind of malware can distort a section of an algorithm and limit its functionality, which creates unreliable behaviour. Logical bombs can either be *syntactic* or *semantic*. The purpose of syntactic bombs is to destroy the logic of the information system by delaying information and/or by developing unpredictable behaviour through malware such as viruses and trojans. With semantic bombs the goal is to manipulate information in order to destroy the trust of the system by changing the contents. This influences the opponent's decision-making process.

One problem with cyber attacks, compared to kinetic attacks, is that they can evolve in undesired ways. There are major risks for ripple effects and the "wrong" target being attacked. A future possibility is to develop functionality for *reversible* attacks. This means that the attacker can remote restore the systems to their original status if the operations do not develop as planned.[331]

One can also imagine cases where one party deliberately deceives the opponent that he has been attacked, whereas in fact nothing has happened. The tactic is called *victim illusionary damage*. Through this stratagem the adversary has to put a lot of resources into controlling and securing his systems. This is a cost effective way to create anxiety and undermine the opponent's trust in his own system and information.[332] From an information operation point of view this is attractive. The deceiver is inside the opponent's decision-making process, influencing his thinking.

---

331 Rowe, N. (2010) "Towards Reversible Cyberattacks". U.S. Naval Postgraduate School, Monterey, USA. 9th European Conference on Information Warfare and Security, July 2010, Thessaloniki, Greece.
332 Rowe, N. (2010).

Psychological influences

Always being connected to the Internet creates new conditions for psychological influences. From a security policy point of view, cyber influences are an important method to transform power relations between different parties and can be a way of getting "soft power".[333]

Cyberspace is neither homogenous nor diametrical; it is a hybrid environment of interactions between people from the all over world with different opinions and ideas; background knowledge, motives and ambitions.

In contrast to the Cold War structure, where mass communication via TV and radio was one-way and top-down, present and even more future communications will be multi-dimensional and bottom-up. Information continuously circles the global network, where it is transformed and changed. This makes it difficult to control and steer information, since it can pass between different access forms and distribution channels more or less uncontrolled and in real time. This, of course, is on condition that no serious restrictions are introduced against use of the Internet, or that the system for some reason is shut down, wholly or partly.

Digitisation of information means that more or less all text, speech, images and sounds created or depicted by humans can be transmitted. Increased processing power, new display techniques etc. result in difficulties when it comes to separating animated, created pictures from real ones. Digital reproductions – avatars – can be made humane with a made-up history and background, a.k.a. "false legends".[334] This opens up for deception of the rest of the world. Along with this, software is being developed that can translate from one language to another, more or less in real time. Avatars and advanced translation programs in combination will give a good potential for influencing operations.

It is not improbable that in the near future people will suffer physical-virtual thefts, not only of personal information, but also of entire identities and "lives" on the Internet, something like today's face rapes on Facebook – but on a larger scale. In cyber crime, identity theft is not an unusual phenomenon. Based on the hijacked identity, avatars are constructed that look

---

333 Kramer, F., Wentz, L. (2008) "Cyber Influence and International Security". Defense Horizons. Center for Technology and National Security Policy, National Defence University. January 2008.
334 Williams, J. (2010) "How to create human avatars". 11 May 2010. Internet article.

like the person that has lost his identity. This kind of false alias can be used by criminals for blackmail and cyber espionage.

Digitisation technology is cheap, and the cost will sink even more. Principally, a good computer with access to the Internet and a camera is all that is needed. That means the knowledge will be passed on to more people. Just as images can be digitized, sounds can be converted into ones and zeros; recorded, analysed and manipulated. From a legal perspective, the development is likely to cause severe problems in view of the burden of proof.

Digitisation paves the way for deception operations on a grand scale. Using holographic 3D images amplifies the message, since more senses are involved. The development towards an all-embracing global network means that hundreds of millions of people can be approached, simply and at a low cost. As tools for relations marketing are refined, this kind of knowledge can also be applied to psychological operations.

As services are individualised, the future Internet will convey unique messages to every individual who is going to be influenced. As opposed to previous phases, mass communications will be individualised and selective.

One effect of increased processor power in computer systems is that individually adapted messages can be monitored, changed and refined automatically. Attitude research and surveys can be made in real time on the Internet in order to analyse how well the messages are received. In traffic analysis, attitudes are measured instantly. If necessary, that is if the messages do not have the intended effect, the direction of the operation is changed. Naturally, this will be done without the manipulated party being aware of it. It is in the nature of deception campaigns. The person who misleads wants to avoid detection, and the person who is being misled does not even know it until it is too late.

To conclude; everything that is written on websites, Internet forums, YouTube etc. is not necessarily true, or correct, it is not even likely. That is why it is important always to be critical. Examples of questions that should be asked by each individual include: Where does the information come from, and why now? What kind of message does it convey, and to whom? For what purpose can it be used and who actually benefits from it?

Conclusions

Let's remind ourselves of one of Sun Tzu's stratagems. In order to create your own strategies and attitudes, you have to know your enemy and what methods and tricks he might use. During the Cold War you knew who the enemy was, you knew his capacity, but not when an attack could occur. In the age of information warfare and cyber threats it is difficult to know who the enemy is, what kind of resources he might have and his underlying motives.

There is a struggle for information and the right to use it. Using the theories of von Clausewitz on the friction of war, the cyber mist is getting thicker. Information can be intercepted, disrupted, manipulated, distorted and destroyed, for a certain purpose or somebody's intentions and beliefs. And it is being done.

This book describes a number of phenomena of more or less serious kind, which occur and increase in magnitude because of the information technological revolution. The paradigmatic shift, as of the origin of a new logic and new kinds of behaviour, has a fundamental impact on societal structures. The world is being connected into an enormous net of networks where information can be disseminated to everybody. We are just in the early phases of the process. Basically the change is very positive, but as with everything it also brings new kinds of risks. This means new challenges to handle.

In order to reduce threats and vulnerabilities when it comes to malicious activities in cyberspace, organisations have to be able to protect themselves and secure information and information systems against threats that cover the whole spectrum. This requires adaptive strategies and attitudes. However, concentrations on security must not be one dimensional, but parallel and in concord with what is the strength and characteristic of the Internet; its openness and accessibility. The one is a condition for the other. Without a certain level of security, it is hard to achieve openness and vice versa. Aspects such as integrity and protection of individuals also have to be remembered. This is a matter of democracy; basically it is a question of what kind of society we want to live in.

The ambition of this book is to provide an increased understanding of and new insights into a mega trend in progress. Hopefully, the text can serve as a basis for continued discussions on the positive and negative effects of the burgeoning information society. The subject is probably interesting to most people. Through our use of information technology, we are all part of cyber space, with everything that involves.

References

Printed sources

Ackerman, R. (2009) "Threats Imperil the Entire U.S. Infrastructure. From the military to the economy, the country is open to vast damage". *SIGNAL, AFCEA,* International Journal, July 2009.

Alberts, D.S, Hayes, R.E. (2003) "Power to the Edge. Command, Control in the Information Age". DoD Command and Control Research Program, *CCRP* Publication Series.

Alberts, D., Gartska, J., Stein, F. (1999) "Network Centric Warfare: Developing and leveraging information superiority". *CCRP Publication Series.* (2:nd edition).

"Antennen gedreht". Wirtschaftwoche no 46, 9 November 2000. Published in the European Parliament Report 2001.

"Annerkennungen zur sicherheitslage der Deutschen Wirtschaft". ASW; Bonn, April 2001, published in the European Parliament Report 2001.

Atwan, A.B. (2006) "The secret History of al Qaeda". *University of California Press,* Berkley, USA

Augustsson, T. (2010) "Litet hopp stoppa nätvirus". *Svenska Dagbladet,* 16 June 2010.

Barret, B.M. (2008) "Information Warfare: China's response to U.S. technological Advantages". International Journal of Intelligence, vol. 18, no 4.

Bishop, M. (2003) "Computer Security Art and Science". *Pearson Education Inc.,* New Jersey, 2003.

Bogdanov, S.A. (2004) "Warfare of the Future". *Military Thought,* vol. 13, no 1.

Bowcott, Owen. (2009) "Abdul Qadeer Khan. Pakistani nuclear scientist accused of industrial espionage". *The Guardian,* 6 February 2009.

Brenner, B. (2007) "Black Hat 2007: Estonian attacks were a cyberriot, not warfare". Information Security Magazine, 3 August 2007.

Carr, J (2010) "Inside Cyber Warfare. Mapping the Cyber Underworld". *O'Reilly Media Inc.,* Sebastopol CA, USA.

Carman, D. (2002) "Translation and Analysis of the Doctrine of Information Security of the Russian Federation: Mass media and the politics of identity". *Pacific Rim Law & Policy Journal Association.*
Cervenka, A. (2010) "Spionjägare". *Svenska Dagbladet,* 18 July 2010.
Christiansson, H., Fischer, G. (2003) "Terrorismens tid". *SNS Förlag.*
Claburn, T. (2009) "Heartland Payment System Hit By Data Security Breach". *Information Week,* 20 January 2009.
Clausewitz, C. von (1832). "Om kriget". Translation made by Hjalmar Mårtensson, Klaus-Richard Böhme, Alf W. Johansson. New print, *Bonniers AB,* Stockholm 2002.
Chies, R., Ducci, S, Ciappi, S. (2009) "Profiling Hackers. The Science of Criminal Profiling as Applied to the World of Hacking". *CRC Press Taylor & Francis Group,* NW, USA.
"China's Restless Hackers". *Strategypage,* Information Warfare Article Index, 18 April 2006.
"Chinese 'patriotic' hackers billed as network security research organization". *East-Asia-Intel Reports,* 4 December 2006.
Chowdhuri, S.R. (2002) "Subversive activities through cyber space". *The Tribune,* 20 May 2002.
Cole, M. (2009) "Friendship is no bar to espionage". MOSS. *MacArthur Centre for Security Studies,* 1 November 2009.
Collin, B. (1997) "The future of Cyberterrorism: Where the Physical and Virtual Worlds Converge". 11th Annual International Symposium on Criminal Justice Issues. The University of Illinois at Chicago.
CSR Report for Congress. "Botnets, Cyber crime and Cyber terrorism: Vulnerabilities and Policy Issues for Congress". Updated 29 January 2008.
CRS Report for Congress. "Cyberwarfare". Updated 19 June 2001.
"Dags för ett nytt RMA". FOI Framsyn no 2, 2001.
Danielsson, L. (2010) "Räkna med spioner i företaget". *Computer Sweden,* 23 March 2010.
Dennings, D. (2000) "Cyberterrorism". Testimony before the Special Oversight Panel on Terrorism Committee on Armed Services, US House of Representatives, 23 May 2000.
Dong-A, I. (2007) "China Wants Dominance in Cyber Space". *ROK Daily,* 21 September 2007.
Donskov, Y., Nikitin, O.G. (2005) "Special Information Operations in Armed Conflicts". *Military Thought,* vol. 14, no 3.

Dylevsky, I.N., Komov, S.A., Korotkov, S.V., Rodionov, S.N., Fedorov, A.V. (2007) "Russian Federation Military Policy in the Area of International Information Security: Regional aspect". *Moscow Military Thought*, 31 March 2007.

European Parliament Report 2001 on the existence of global system for the interception of private and commercial communications (ECHELON interception system) (2001/2098(INI)) RR\445698EN.doc

Fitzgerald, M. (1996) "Russian Views on Information Warfare". December 1996. Hudson Institute. Washington D.C, USA

Fylkner, M, Carlsen, H., Lewrentz, B. (2004) "Aktörer, antagonister och angrepp. En studie om det kvalificerade IT-hotet". Användarrapport FOI.

Förster, A. (2001) "Maulwurfe in Nadelstreissen". Published in the European Parliament Report 2001.

Giacomello, C. (2004) "Bangs for the Buck: A Cost-Benefit Analysis of Cyberterrorism". Studies in Conflict & Terrorism. University of Bologna, Italy. *Taylor & Francis Group*.

Goldberg, D. (2007) "Handelsbanken drar tillbaka kortnummer". *Computer Sweden*, 7 April 2007.

Goldberg, D., Larsson, L. (2007) "Banker tiger om stulna kortnummer". *Computer Sweden*, 20 June 2007.

Gorman, S., Cole, A., Dreazen, Y. (2009) "Computer Spies Breach Fighter-Jet Project". *The New York Times Journal*, 21 April 2009.

Grau. L.W., Thomas, T. (1996) "A Russian View of Future War: Theory and Direction". *Journal of Slavic Military Studies*, issue 9.3, September 1996.

Greenemeir, L. (2007) "Electronic Jihad App Offers Cyberterrorism for the Masses". *Information Week*, 2 July 2007.

Handley, L. (2007) "Intel Brief: Afghan Poppies Fear Not". Center for Security Studies and Conflict Research, Swiss Federal Institute of Technology, 2007.

Hart, K. (2008) "Longtime Battle Lines Are Recast in Russia and Georgia's Cyberwar". *The Washington Post*, 14 August 2008.

Heickerö, R. (2010) "Cyberkrig mellan hackergrupper utvecklar nya hotbilder". In Granholm, N., Lindström, M. (editor). 'Strategisk utblick 2010: Säkerhetspolitisk nattorientering?' Användarrapport FOI, June 2010.

– (2010) "Cyber security challenges for Asia in a 2030 time frame". Asian Security Conference 2030, IDSA, New Delhi, 11–13 February 2010.

- (2010) "Emerging Cyber Threats and Russian Views on Information Warfare and Information Operations". Användarrapport FOI, Mars 2010.
- (2009) "Cyberkriget eskalerar och utvecklar nya hotbilder". In Granholm, N., Malminen, J., Rydqvist, J. (editor) "Strategisk utblick 2009. Säkerhetspolitisk forsränning?" Användarrapport FOI, April 2009.
- (2008) "Terrorism online and the change of modus operandi". 13th ICCRTS Symposium, Seattle, USA, 17–19 June 2008.

Heickerö, R., Hyberg, P., Bäckström, M., Olsson, G., Renhorn, I., Jonason, T., Eklöf, F. (2004) "Telekrig i en breddad hotbild". Underlagsrapport FOI, December 2004.

Heickerö, R., Larsson D. (2008) "Terror Online. Cyberhot och Informationskrigföring". *Conopsis Förlag*, Stockholm.

Hoffman, D. (2008) "KGB Comes in from the Cold". *The Washington Post*, 8 December, 2008.

Hoofnagle, C. (2007) "Identity Theft: Making the Known Unknowns Known". *Harvard Journal of Law and Technology*, vol. 21, 2007 (autumn).

"ICT and International Security". Disarmament Forum, no 3. 2007

"Indisk konflikt kan gynna Ericsson". *Dagens Industri*, 15 May 2009.

InfoSecurity. "Grey Goose 2 Ties Kremlin More Closely to Georgia Cyberattacks". *Info Security Magazine*, 20 March 2009.

Jiangzhou, L., Dehui, X. (2000) "Planning and Application of Strategies of Information Operations in High tech Local War". *Zhongguo Junshi kexue (China Military Science)*, no 4, 20 August 2000.

Johnson, B. (2009) "No One Is Ready for This". *The Guardian*, 16 April 2009.

Kash, W. (2008) "Lessons from the Cyberattacks on Estonia. Interview with Lauri Allman, Estonia's permanent undersecretary of defense". *The Government Computer News*, 13 June 2008.

Keggler, J. (2008) "Taking the Fight to the Net". *Armada International*, vol. 32, issue 2 (April/May).

Komov, S.A., Korotkov, S.V., Rodionov, S.N. (2003) "International Information Security: Military aspects". *Military Thought*, vol. 12, no 4.

Korotchenko, Y., Plotnikov, N. (1994) "Information is also a Weapon: About what should not be forgotten when working with personnel". *Krasnaya Zvezda*, 17 February 1994.

Kramer, F., Wentz, L. (2008) "Cyber Influence and International Security". *Defense Horizons*. Center for Technology and National Security Policy, National Defense University. January 2008.
Krebs, B. (2007) "Security Fix. Calculating the Costs of Cyber Crime". *The Washington Post*, 27 September 2007.
Krishnan, N.R. (2010) "Defence R&D in Asia: Achievements & Future Directions. 2010. Asian Militaries and the Future of War". IDSA Conference, New Delhi, 12 February 2010.
Kukashkin, A.N., Yefimov, A.I. (1995) "The Security of the Infosphere of Strategic Defence Systems". *Military Thought*, no 5.
Leijonhielm, J., Hedenskog, J., Knoph, T., Larsson, R., Oldberg, I., Roffey, R., Tisell, M., Westerlund, F. (2008) "Rysk militär förmåga i ett tioårsperspektiv – ambitioner och utmaningar". Användarrapport FOI.
Leijonhielm, J., Hedenskog, J., Knoph, J., Oldberg, I., Unge, W., Vendil, C. (2000) "Rysk militär förmåga i ett tioårsperspektiv. En förnyad bedömning 2000". Användarrapport FOA.
Li, Y. (1996) "New Subjects of study brought about by Information Warfare". Jiefangjun bao,1996.
Libicki, M. (2009) "Cyberdeterrence and Cyberwar". *RAND Corporation*, Santa Monica, CA, USA.
Limno, A.N., Krysanov, M.F. (2003) "Information Warfare and Camouflage, Concealment and Deception". *Military Thought*, vol. 12, no 2.
Lynn, W. (2010) "Defending a new Domain". *Foreign Affairs*. Pepperdine University. School of Public Policy. September 2010.
MacCrory, D. (2000) "UK Muslims Volunteers for Kashmir War'" *The Times*, 28 December 2000.
Manwaring, M. (2006) "The Asymmetrical 21st Century Security Reality". Strategic Studies Institute. US Army War College. *IOS Press,* 2006.
Markoff, J. (2009) "Defying Experts, Rogue Computer Code Still Lurks". *The New York Times*, 28 August 2009.
– (2008) "Before the Gunfire, Cyber attacks". *The International Herald Tribune*, 13 August 2008.
– (2009) "Vast Spy System Loots Computers in 103 Countries". *The New York Times*, 28 March 2009.
– (2009) "Chinese hackers using ghost network to control embassy Computers". *The Times*, 29 March 2009.
– (2008) "Before the Gunfire, Cyberattacks". *The International Herald Tribune*, 13 August 2008.

Markoff, J., Kramer, A. (2009) "U.S. and Russia Differ on Treaty for Cyberspace". *The New York Times*, 27 June, 2009.
Matthews, W. (2008) "New Ways of War: Cyber attacks likely in any military Conflict". *Defence News Reports*, 26 October 2008.
McAfee Virtual Criminology Report. (2007) "Cybercrime: The next wave, 2007".
Meltzer, J. (2010) "SAS har dömts för industrispionage i Norge". *Affärsvärlden*, 7 March 2010.
Mowthorpe, M. (2005) "The Revolution in Military Affairs (RMA): The United States, Russian and Chinese Views". University of Hull, vol. 5, no 2 (summer)
Napoleoni, L. (2004) "Oheligt krig. Den moderna terrorismens ekonomiska rötter". Andersson Pocket AB, Stockholm.
Nato (2009) "Multiple Futures Project. Navigating towards 2030". Final report, April 2009.
"New Symantec Report Reveals Booming Underground Economy". *Information Systems Security*, 24 November 2008.
Nicoll, A. (editor). "Chinas cyber attacks. Casting a wider intelligence net". *IISS Strategic Comments*, vol. 13 issue, 7 September 2007.
Nolen, S. (2005) "Nigerian e-mail scammers feeding on greed, gullibility". *The Globe and Mail*, 5 December 2005.
Noonan, S. (2010) "China and its Double-edged Cyber-sword". *Stratfor*, 9 December 2010.
"Nätverksbaserat Försvar. Fördjupade studier utifrån Försvarshögskolans Sakområden". Swedish National Defence College, 2003.
Osipovich, A. (2007) "Inside a Hacker school". *Foreign Policy*, issue 163, Nov/Dec 2007.
Owens, W. (1996) "The Emerging U.S. System-of-Systems". National Defense University, USA.
Pirumov, V. (1996) "Nekotorye aspekty informatsionnoi voiny (Certain aspects of information warfare)". Conference speech in Moscow, May 1996.
Pollard, N. (2005) "UN report puts world's illicit drug trade at estimated $321". *The Boston Globe Journal*, 30 September 2005.
Porthillo-Shrimpton, T. (2008) "Battle for South Ossetia Fought in Cyberspace". *The Independent*, 17 August 2008.
Rastorguyev, S.G. (1998) "Informatsionnoi Voiny (Information warfare)". *Radio Svjaz*.

Rowe, N. (2010) "Towards Reversible Cyberattacks". U.S. Naval Postgraduate School, Monterey, USA. *9th European Conference on Information Warfare and Security*, Thessaloniki, Greece, July 2010.
Roy, B. (2009) "China's Silent warfare". *South Asia Analysis*, paper no 3147, 13 April 2009.
Schecter, E (2008) "U.S. struggles to break ahead of rivals in network security race". *C4ISR Journal*, 1 March 2008.
Schütze, von A. (1998) "Wirtschaftspionage: Was macht eigentliche die Konkurrenz?". No 1, 1998, published in the European Parliament Report 2001.
Serookiy, Y. (2004) "Psychological-Information Warfare: Lessons of Afghanistan". *Military Thought*, vol. 13, no 1.
Shactman, N. (2009) "Kremlin Kids: We Launched the Estonian Cyber War". *Wired*, 11 March 2009.
Shah, A. (2009) "World Military Spending". *Global Issues*, 7 July 2010.
Simmons, C. (2006) "The Huawei Way". *Newsweek*, 16 January 2006.
SIPRI. (2011) "The Financial value of global arms trade". 11 October 2011.
"Some Experts Question Activities of China's Patriotic" Hackers". *World News Connection*, 10 July 2007.
Stallin, W. (2003) "Network Security Essentials. Applications and Standards". *Pearson Education Inc.*, New Jersey, (2nd Editions).
Stoll, C. (1990) "The Cuckoo's Egg: Tracking a Spy Through the Maze of Computer Espionage". *Pocket Books*.
Symantec. "Internet Security Threat". Report 2007.
Thomas, T. (2004) "Russian and Chinese Information Warfare: Theory and Practise". Foreign Military Studies Office, Fort Leavenworth. PowerPoint, June 2004.
– (2004) "Chinese Information Warfare Theory and Practice". Foreign Military Studies Office (FMSO), Fort Leavenworth, Kansas. April 2003.
– (2003) "Manipulating the Mass Consciousness: Russian & Chechen Information War. Tactics in the second Chechen–Russian conflict". Foreign Military Studies Office, Fort Leavenworth, Kansas, April 2003.
– (2003) "Al Qaeda & the Internet: the Dangers of Cyber planning". Foreign Military Studies Office, Fort Leavenworth, Kansas 2003.
– (1998a) "Dialectical versus Empirical Thinking: Ten key elements of Russian understanding of information operations". FMSO Special Study Center For Army Lesson Learned. Fort Leavenworth, KS 66027–1327.

- (1998b) "Russia's Information Warfare Structure: Understanding the roles of the Security Council, FAPSI, the State Technical Commission and the military". *European Security*, vol. 7, no 1 (spring).
- (1996) "Russian Views on Information-Based Warfare". *Airpower Journal*, July 1995.

Tkacic, J. (2008) "Trojan dragon: China's Cyber Threats". *The Heritage Foundation*, Washington DC, USA, 8 February 2008.

Traynor, I. (2007) "Russia Accused of Unleashing Cyberwar to Disable Estonia". *The Guardian,* 17 May 2007.

Triplett, W.C. (2002) "Potential Applications of PLA Information Warfare Capabilities to Critical Infrastructures". *Hampton Roads International Security Quarterly.*

"Tsai i Chih Chung: Suntzi talar, Krigskonsten". *Alhambras Förlag* 1996.

Tsymbal, V.I. (1995) "Kontseptsiya Informatsionnoi Voiny (Concepts of Information Warfare)". Presentation at the Russian–U.S. conference on Evolving Post Cold War National Security Issues, Moscow, 12–14 September.

United Nations, Security Council Committee established pursuant to resolution (1267 81999), 20 September 2002.

US-CCU (2009) "Overview by the US-CCU of the Cyber Campaign against Georgia in August of 2008". Special report, August 2009.

Vause, J. (2008) "No site is safe". CNN.com, 7 March, 2008.

Vendil Pallin, C., Westerlund, F. (2010) "Russia's Military Doctrine – Expected News". *FOI RUFS Briefing no 3*, February 2010.

Verisign. (2008) "The Russian Business Network: Rise and fall of criminal ISP". IDefence Security Report, 8 March 2008.

Wang, B. (1997) "A preliminary Analysis of IW". *China Military Science*, no 4, 20 November 1997.

Waterman, S. (2008) "Analysis: Russia-Georgia Cyberwar Doubted". *United Press International*, 14 August 2008.

Watzlawick, P. (1976) "How real is real? Confusion, disinformation, communication". *Random House*, New York, USA.

Weinmann, G (2006) "Terror on the Internet: the new arena challenges". U.S. Institute of Peace. Washington D.C. USA.

Wen, T. (2002) "PLA Bent on Seizing Information Control". *Hong Kong Ching Pao*, 1 June 2002.

Wright, L. (2007) "Al-Qaida och vägen till 11 september". *Albert Bonniers Förlag.*

Young, C. (2007) "Putin's Young Brown Shirts". *The Boston Globe*, 10 August 2007.
Zeller, J. (2006) "Cyber thieves Silently Copy Your Passwords as You Type". *The New York Times*, 27 February 2006.

Articles on the internet

Adhikari, R. (2008) "RBS WorldPay Data Breach Hits 1.5 Million". *Internetnews*, 24 December 2008.
http://internetnews./com/security/article.php/3793386/RBS-WorldPay-Data-Brech-Hits-15-Million.htm.
"Afghan Taliban website hacked as Internet outlawed". 27 August 2001.
http://www.hackinthebox.org/modules.php?
Beaumont, C. (2010) "Stuxnet virus: worm could be aimed at high-profile Iranian Targets". *The Telegraph*, 23 September 2010.
http://www.telegraph.co.uk/technology/news/8021102/Stuxnet-virus-worm-could-be-aimed-at-high-profile-iranian-target.html.
Beichmann, A. (1993) "Reagan ought to get Oscar for Star Wars – how former President Ronald Reagan used the threat of proposed Strategic Defense Initiative to weaken the Soviet Union".
Column. *News World Communication Inc.*
http.//findarticles.com/p/articles/mi_m1571/is-n39-v9/ai_14277739/
Borger, Julian (2002) "US fears al Qaida hackers will hit vital computer networks". *The Guardian*, 28 June 2002.
http://www.guradian.co.uk/technology/2002/june/28/security.hacking
"BSNL gets nod to award to Huawei". *India Times*.
http://economictimes.india-times.com/News-by-Industry/BSNL-gets-nod-to-offer-deal-toHuawi/articleshow/4537167.cms
Carr, J. (2009) 27 July 2009. http://intelfusion.net/wordpress/?tag=russia
"Chinese Espionage: Britain's MI5 reports epidemic in spying". *Examiner.com*.
http://www.examiner.com/x-2684-Law-Enforcement-Examiner~y2009 m8d17-Chinese-Espionage-Britains-MI5-reports-epidemic-in-spying
Cochran, A. (2006) "Muslim Hackers Assaulting Websites Since Cartoon Controversy Began".
http://counterterrorismblog.org/2006/02/muslim_hackers_assaulting webs.php

Cruickshank, P (2010) "U.S. citizen believed to be writing for al Qaeda website, source says". *CNN News US*, 18 July 2010. http://articles.cnn.com/2010-07-18/us/al.qaeda.magazine_1_qaeda-yemeni-americanlaw-enforcement?_s=PM:US

"Cybercrime Kingpin Indicted". *The New New Internet*. http://.thenewnewinternet.com/2009/08/18/cybercrime-kingpin-indicted/

"Defectors say China running 1,000 spies in Canada". *CBC Canada*. http://www.cbc.ca/canada/story/2005/06/15/spies050615.html#ixzz0xbZs8G4c

"Doktrina Informatsionnoi Bezopasnosti Rossiiskoi Federatsii". http://www.scrf.gov.ro/Documents/Decree/2000/09-09.html

"FBI Computer Crime Survey 2005". www.fbi.gov/publications/ccs2005.pdf

Fitzgerald, M. (1994) "Russian Views on Electronic Warfare. The growing role of information technology is rapidly lowering the barrier between war and peace". Powerpoint, http://www.nationalstrategies.com

Flook, K. (2009) "Russia and the Cyberthreats". 13 May 2009. http://www.criticalthreats.org/russia/russia-and-cyberthreat.

Francis, B. (2005) "Hackers sell their information anonymously through secretive websites. Know thy hacker". *Infoworld*, 28 January 2005. http://www.infoworld.com/article/05/28/05OPPsecadvise_11.html

Garthwaite, J. (2010) "Hybrid Espionage! GM takes couple to court over tech secrets". *Erat2tech*, 23 July 2010. http://gigaom.com/cleantech/hybrid-espionage-gm-takescouple-to-court-over-tech-secrets/

Graham, J. (2000) "Military Power vs Economic Power in History". http://www.historyorb.com/world/power.shtml

"Hackare ville förhandla med Canal Digital". 19 October 2010. http://www.dn.se/nyheter/sverige/hackare-ville-forhandla-med-canal-digital-1.1192221

"Huawei in Spying Flap". 24 June 2004. http://www.lightreading.com/document.asp?doc_id=55172

IntelliBriefs. "Project Grey Goose: Some lessons learned, and more to come". 22 October 2008. http://intellibriefs.blogspot.com/2008/10/project-grey-goose-some-lessonslearned

"IT & Software opportunities in Moscow". Moscow Investment Gateway. Moscow Government; Goskomstat; Statistics on Russian Education. *Watson Wyatt.*
http://74.125.77.132/search?q=cache:ktOSEdAuIOcJ:moscow.eregula tions.org/Media/Editor_Repo/undp_it%2520%26%2520software.ppt+ unemployment+rate+ict+sector+russia&cd=2&hl=sv&ct=clnk&gl=se
Jamshed, A. (2008) "Cyberwars between India and Pakistan". http://www.Goarticles.com/cgi-bin/showa.cgi?C=859977
Knapp, M. (2002) "The War of the Ether: Al-Qaeda's PSYOPS campaign against the Western & Muslim Worlds". http://www.alansar.biz
Lettice, J. (2007) "Jailed terror student "hid" files in the wrong Windows folder and provided terror instruction via web links".
The Register, 23 October 2007.
http://theregister.co.uk/2007/10/23/siddique_trial_sentencing/print.html.
MacQuaid, J. (2008) "The RBN Operatives Who Attacked Georgia Secure Home Network". 18 August 2008.
http://securehomenetwork.blogspot.com/2008/08/rbn-operatives-who-attackedgeorgia.html
Marturion, D. "Will hackers keep the cyber peace?". *New Business News.*
http://www.newbusinessnews.com/story/11160101.html
Mills, E. (2009) "Cybercrime cost firms $1 trillion globally".
CNET News, 28 January 2009.
http://news.cnet.com/8301-1009_3-10152246-83.html?tag=leftCol; post-4438
"Moe funnet död". *Dagbladet.*
http://www.vg.no/nyheter/innenriks/artikkel.php?artid=2885854
"Nigeriabrev. 419 Scam". *Antarktis media*, 2002-2004.
http.//www4.tripnet.se/-pingo/419/faq.html
O'Connell, K. (2007) "Russian company outed as mother of all cyber-crime". *INTERNET LAW*, 24 October 2007.
http://www.ibls.com/internet_law_news_portal_view.aspx?id=1887& s=latestnews
"Online Russian blackmail gang jailed for extorting 4 million USD from gambling Websites". *Sophos*, 5 October 2006.
http://www.sophos.com/pressoffice/news/articles/2006/10/extort-ddosblackmail.Html

"Project Grey Goose Phase II Report. The evolving state of cyber warfare". 19 March 2009.
http://www.scribd.com/doc/13442963/Project-Grey-Goose-Phase-II-Report
Raza, M.A. (2008) "India-Pakistan in State of Cyber War". 24 November 2008. http://www.propakistani.com/2008/11724/here-we-go-again/
Richard, J. (2008) "Number of Computer Viruses Tops One Million". *The Times online*, 10 April 2008.
http://technology.timesonline.co.uk/tol/news/tech_and_web/article3721556.ece
Rogin, J. (2006) "China fielding cyberattacks units".
http://www.fcw.com/online/news/94650-1.html
"ROK Monthly Claims DPRK hacks, Steals Military secrets in March 2009". *Chosun Ilbo Online*, 19 October 2009.
http://dlib.eastview.com/browse/doc/20797846
"Russia Employment Rate. Trading economic. Global Economics Research".
http://www.tradingeconomics.com/Economics/Unemploymentrate.aspx?symbol=RUB
"Russian Business Network (RBN)". October 2007.
http://rbnexploit.blogspot.com/2007_09_01_archive.html
"Russian Federation Law No, 40-ZFZ. On Organs of the Federal Security Service in the Russian Federation".
http://fas.org/irp/world/russia/docs/law_950403.htm
"Smarta högutbildade offer för aktiebluff".
http://www.dagensps.se/artiklar/tt/2010/09/21/aktiebluff/index.xml
Sokov, N. (2004) "Russia's 2000 Military Doctrine". July 2004
http://nti.org/dbnisprofs/over/doctrine.htm
"Spam kostar miljarder". *IDG.se*. 12 March 2004.
http://www.idg.se/2.1085/1.19212
"Spionage 1, 2010". Säkerhetspolisen. 17 May 2010.
http://www.sakerhetspolisen.se/download/18.34ffc68f1235b740c0680001750/spionage12010.pdf
Stevens, K., Jackson, D. (2010) "Zeus Banking Trojan Report". *Secure Works*, 11 March 2010.
http://www.secureworks.com/research/threats/zeus/?threat=zeus
"The Cyber Raiders Hitting Estonia". *BBC*, 17 May 2007.
http://news.bbc.co.uk/2/hi/europe/6665195.stm

"The Dark Visitor". 4 April 2008. http://www.thedarkvisitor.com.
"The Russian Concorde Tupolev TU-14".
http://www.gizmohighway.com/history/tu-144.htm
Thomas, T. (1996) "Deterring Information Warfare: A new strategic challenge". *IWS* – the Information Warfare Site.
http://www.iwar.org.uk/iwar/resources/parameters/iw-deterrence.htm
"Towards 50 Billion devices connected?".
http://dw2blog.com/category/connectivity
United Press International (2007) "Analysis: Who Cyber Smacked Estonia?".
Phttp://www.upi.com/security/security_terrorism/Analysis/2007/06/analysis_who_cybersmacket_estonia/2683/print_view/
"US Airforce Seeks Cyber Security in a Cloud". *DefenceIQ*. 26 April 2010.
http://www.defenceiq.com/article.cfm?externalid=2320
Wang, B. (2005) "The challenge of Information Warfare". *FAS*.
http://www.fas.org/irp/world/china/docs/iw_mg_wang.htm
Wang B, Li, F. (1995) "Information Warfare". *FAS*.
http://www.fas.org/irp/world/china/docs/iw_wang.htm
Weimann, G. (2004) "WWW.Terror.Net: How terrorism uses the Internet". United States Institute for Peace.
www.usip.org/pubs/specialreports/sr116pdf.
Weinberg, N. (2007) "Kasperskys on cybercrime: Don't blame the Russian Maffia and why we need anti-anti-anti virus software".
Network World, 2 January 2007.
http://www.networkworld.com/news/2007/020107-kaspersky-cybercrime.html
Williams, J. (2010) "How to create human avatars". 11 March 2010.
http://www.ehow.com/how_6501921_create-human-avatars.html
WMD Insights. "Recent Events Suggest Cyber Warfare Can Become New Threat". December 2008/January 2009.
http://wmdinsights.com/129/129_G3_recentEvents.htm
"Voyennaia Doktrina Rossiiskoy Federatsii". Utverzhdena Ukazom Prezidenta RF ot 21 aprelya 2000 g. no 706.
http://www.scrf.gov.ru//Documents/Decree/2000/706-1.html.
"2007 Cyber Attack on Estonia Launched by Kremlin-backed Youth Group". *Homeland Security News Wire*.
http://homelanssecuritynewswire.com/2007-cyber-attack-estonia-launched-kremlinbacked-youth-group

http://attribute.org/news/content/01-10-22001.html
http://economictimes.indiatimes.com/News-by-Industry/BSNL-gets-nod-to-offerdeal-to-Huawei/articleshow/4537167.cms
http://intellibriefs.blogspot.com/2009/03/cyber-warfare-project-grey-goosephase.html
http://internet-haganah.com/haganah/index.html
http://wapedia.mobi/nashi_(youth_movement)
http://www.cbc.ca/canada/story/2005/06/15/spies050615.html
http://www.cert-in.org.in/knowledgebase/whitepapers/analysis_deface websites.htm
http://www.danielpipes.org/rr/1659.php
http://www.dtic.mil/jv2010/jv2010.pdf
http://www.globalterroralert.com/azzam-mokhtar.pdf
http://www.guardian.co.uk/world/2009/feb/06/pakistani-scientist-abdul-qadeer-khan
http://www.infowar-monitor.net/
http://www.islamicawakening.com/authors.php?authorlist=121&
http://www.jehad.net
http://www.jihadunspun.net
http://www.rferl.org/Content/Behind_The_Estonia_cyberattacks/1505613.html
http://www.scribd.com/doc/2196587/Cyber-Warefare
http://www.uk/cip/resources/nipc/cyberprotestupdate
http://www.usip.org/pubs/specialreports/sr116pdf

Other references

The Economist, volume 384, no 8545, 8 September 2007.
"Trojan bakom flygkrasch". *TT*, 21 August 2010.
Robert Lyle, Radio Liberty/Radio Free Europe, 10 February 1999.

www.ingramcontent.com/pod-product-compliance
Ingram Content Group UK Ltd.
Pitfield, Milton Keynes, MK11 3LW, UK
UKHW021830140426
5217IPUK00021B/1360